A FRIENDLY APPROACH TO
COMPLEX ANALYSIS

A FRIENDLY APPROACH TO
COMPLEX ANALYSIS

Sara Maad Sasane
Lund University, Sweden

Amol Sasane
London School of Economics, UK

 **World Scientific**

NEW JERSEY · LONDON · SINGAPORE · BEIJING · SHANGHAI · HONG KONG · TAIPEI · CHENNAI

Published by

World Scientific Publishing Co. Pte. Ltd.

5 Toh Tuck Link, Singapore 596224

USA office: 27 Warren Street, Suite 401-402, Hackensack, NJ 07601

UK office: 57 Shelton Street, Covent Garden, London WC2H 9HE

British Library Cataloguing-in-Publication Data
A catalogue record for this book is available from the British Library.

Credit for Image from Wikimedia Commons:
Color plot of complex function $(x^{2-1}) * (x-2-I)^2 / (x^2+2+2I)$, hue represents the argument, sat and value represents the modulus (Permission=CC-BY 2.5).
Image creator: Claudio Rocchini
Source: http://en.wikipedia.org/wiki/File:Color_complex_plot.jpg

A FRIENDLY APPROACH TO COMPLEX ANALYSIS

ISBN 978-981-4578-98-1
ISBN 978-981-4578-99-8 (pbk)

Printed in Singapore

Preface

We give an overview of what complex analysis is about and why it is important. As the student must have learnt the notion of a complex number at some point, we will use *that* familiarity in our discussion here. Later on, starting from Chapter 1 onwards, we will start things from scratch again. So the reader should not worry about being lost in this preface!

What is Complex Analysis?

In *real* analysis, one studies (rigorously) calculus in the setting of real numbers. Thus one studies concepts such as the convergence of real sequences, continuity of real-valued functions, differentiation and integration. Based on this, one might guess that in *complex* analysis, one studies similar concepts in the setting of complex numbers. This is partly true, but it turns out that up to the point of studying differentiation, there are no new features in complex analysis as compared to the real analysis counterparts. But the subject of complex analysis departs radically from real analysis when one studies differentiation. Thus, complex analysis is not merely about doing analysis in the setting of complex numbers, but rather, much more specialized:

> Complex analysis is the study of "complex differentiable" functions.

Recall that in real analysis, we say that a function $f : \mathbb{R} \to \mathbb{R}$ is *differentiable at* $x_0 \in \mathbb{R}$ if there exists a real number L such that

$$\lim_{x \to x_0} \frac{f(x) - f(x_0)}{x - x_0} = L,$$

that is, for every $\epsilon > 0$, there is a $\delta > 0$ such that whenever $0 < |x - x_0| < \delta$, there holds that

$$\left| \frac{f(x) - f(x_0)}{x - x_0} - L \right| < \epsilon.$$

In other words, given any distance ϵ, we can make the difference quotient

$$\frac{f(x) - f(x_0)}{x - x_0}$$

lie within a distance of ϵ from the real number L for all x sufficiently close to, but distinct from, x_0.

In the same way, we say that a function $f : \mathbb{C} \to \mathbb{C}$ is *complex differentiable at $z_0 \in \mathbb{C}$* if there exists a complex number L such that

$$\lim_{z \to z_0} \frac{f(z) - f(z_0)}{z - z_0} = L,$$

that is, for every $\epsilon > 0$, there is a $\delta > 0$ such that whenever $0 < |z - z_0| < \delta$, there holds that

$$\left| \frac{f(z) - f(z_0)}{z - z_0} - L \right| < \epsilon.$$

The only change from the previous definition is that now the distances are measured with the *complex* absolute value, and so this is a straightforward looking generalization.

But we will see that this innocent looking generalization is actually quite deep, and the class of complex differentiable functions looks radically different from real differentiable functions. Here is an instance of this.

Example 0.1. Let $f : \mathbb{R} \to \mathbb{R}$ be given by $f(x) = \begin{cases} x^2 & \text{if } x \geq 0, \\ -x^2 & \text{if } x < 0. \end{cases}$

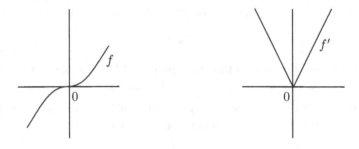

Fig. 0.1   Graphs of the functions f and its derivative f'.

Then f is differentiable everywhere, and

$$f'(x) = \begin{cases} 2x & \text{if } x \geq 0, \\ -2x & \text{if } x < 0. \end{cases} \tag{0.1}$$

Indeed, the above expressions for $f'(x)$ are immediate when $x \neq 0$, and $f'(0) = 0$ can be seen as follows. For $x \neq 0$,

$$\left| \frac{f(x) - f(0)}{x - 0} - 0 \right| = \left| \frac{f(x)}{x} \right| = \frac{|x|^2}{|x|} = |x| = |x - 0|,$$

and so given $\epsilon > 0$, we can take $\delta = \epsilon \ (> 0)$ and then we have that whenever $0 < |x - 0| < \delta$,

$$\left| \frac{f(x) - f(0)}{x - 0} - 0 \right| = |x - 0| < \delta = \epsilon.$$

However, it can be shown that f' is not differentiable at 0; see Exercise 0.1. This is visually obvious since f' has a corner at $x = 0$.

Summarizing, we gave an example of an $f : \mathbb{R} \to \mathbb{R}$, which is differentiable everywhere in $\mathbb{R}$, but whose derivative f' is not differentiable on $\mathbb{R}$.

In contrast, we will later learn that if $F : \mathbb{C} \to \mathbb{C}$ is a complex differentiable function in $\mathbb{C}$, then it is infinitely many times complex differentiable! In particular, its complex derivative F' is also complex differentiable in $\mathbb{C}$. Clearly this is an unexpected result if all we are used to is real analysis. We will later learn that the reason this miracle takes place in complex analysis is that complex differentiability imposes some "rigidity" on the function which enables this phenomenon to occur. We will also see that this rigidity is a consequence of the special geometric meaning of multiplication of complex numbers. $\lozenge$

Exercise 0.1. Prove that $f' : \mathbb{R} \to \mathbb{R}$ given by (0.1) is not differentiable at 0.

Why study complex analysis?

Although it might seem that complex analysis is just an exotic generalization of real analysis, this is not so. Complex analysis is fundamental in all of mathematics. In fact real analysis is actually inseparable with complex analysis, as we shall see, and complex analysis plays an important role in the applied sciences as well. Here is a list of a few reasons to study complex analysis:

(1) **PDEs.** If $f : \mathbb{C} \to \mathbb{C}$ is a complex differentiable function in $\mathbb{C}$, then we have two associated real-valued functions $u, v : \mathbb{R}^2 \to \mathbb{R}$, namely the real and imaginary parts of f: for $(x, y) \in \mathbb{R}^2$, $u(x, y) := \mathrm{Re}(f(x, y))$ and $v(x, y) := \mathrm{Im}(f(x, y))$.

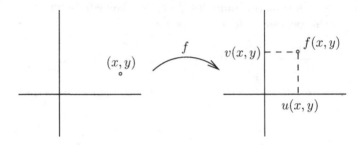

Fig. 0.2   The real and imaginary parts u, v of f.

It turns out that real and imaginary parts u, v satisfy an important basic PDE, called the Laplace equation:

$$\Delta u := \frac{\partial^2 u}{\partial x^2} + \frac{\partial^2 u}{\partial y^2} = 0.$$

Similarly $\Delta v = 0$ in $\mathbb{R}^2$ as well. The Laplace equation itself is important because many problems in applications, for example, in physics, give rise to this equation. It occurs for instance in electrostatics, steady-state heat conduction, incompressible fluid flow, Brownian motion, etc.

(2) **Real analysis.** Using complex analysis, we can calculate some integrals in real analysis, for example

$$\int_{-\infty}^{\infty} \frac{\cos x}{1 + x^2} dx \quad \text{or} \quad \int_{0}^{\infty} \cos(x^2) dx.$$

Note that the problem is set in the reals, but one can solve it using complex analysis.

Moreover, sometimes complex analysis helps to clarify some matters in real analysis. Here is an example of this. Consider

$$f(x) := \frac{1}{1 - x^2}, \quad x \in \mathbb{R} \setminus \{-1, 1\}.$$

Then f has a "singularity" at $x = \pm 1$, by which we mean that it is not defined there. It is, however defined in particular in the interval $(-1, 1)$. The geometric series

$$1 + x^2 + x^4 + x^6 + \ldots$$

converges for $|x^2| < 1$, or equivalently for $|x| < 1$, and we have

$$1 + x^2 + x^4 + x^6 + \cdots = \frac{1}{1 - x^2} = f(x) \text{ for } x \in (-1, 1).$$

From the formula for f, it is not a surprise that the power series representation of the function f is valid only for $x \in (-1, 1)$, since f itself has singularities at $x = 1$ and at $x = -1$. But now let us consider the new function g given by

$$g(x) := \frac{1}{1 + x^2}, \quad x \in \mathbb{R}.$$

The geometric series $1 - x^2 + x^4 - x^6 + - \ldots$ converges for $|-x^2| < 1$, or equivalently for $|x| < 1$, and we have

$$1 - x^2 + x^4 - x^6 + - \cdots = \frac{1}{1 + x^2} = g(x) \text{ for } x \in (-1, 1).$$

So the power series representation of the function g is again valid only for $x \in (-1, 1)$, despite there being no obvious reason from the formula for g for the series to break down at the points $x = -1$ and $x = +1$. The mystery will be resolved later on in this book, and we need to look at the *complex* functions

$$F(z) = \frac{1}{1 - z^2} \text{ and } G(z) = \frac{1}{1 + z^2}$$

(whose restriction to $\mathbb{R}$ are the functions f and g, respectively). In particular, G now has singularities at $z = \pm i$, and we will see that what matters for the power series expansion to be valid is the biggest size of the disk we can consider with center at $z = 0$ which does not contain any singularity of G.

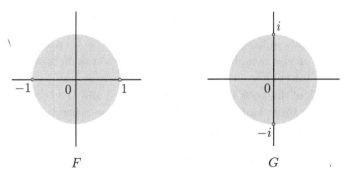

Fig. 0.3   Singularities of F and G.

(3) **Applications.** Many tools used for solving problems in applications, such as the Fourier/Laplace/z-transform, rely on complex function theory. These tools in turn are useful for example to solve differential equations which arise from applications. Complex analysis plays an important in applied subjects such as mathematical physics and engineering, for example in control theory, signal processing and so on.

(4) **Analytic number theory.** Perhaps surprisingly, many questions about the natural numbers can be answered using complex analytic tools. For example, consider the Prime Number Theorem, which gives an asymptotic estimate on the number $\pi(n)$ of primes less than n for large n:

Theorem 0.1. (Prime Number Theorem) $\displaystyle \lim_{n\to\infty} \frac{\pi(n)}{n/(\log n)} = 1.$

It turns out that one can give a proof of the Prime Number Theorem using complex analytic computations with a certain complex differentiable function called the Riemann zeta function. Associated with the Riemann zeta function is also a famous unsolved problem in analystic number theory, namely the Riemann Hypothesis, saying that all the "nontrivial" zeros of the Riemann zeta function lie on the line $\mathrm{Re}(s) = \frac{1}{2}$ in the complex plane. We will meet the Riemann zeta function in Exercise 4.5 later on.

What will we learn in Complex Analysis

The central object of study in this course will be

holomorphic functions in a domain

that is, complex differentiable functions $f : D \to \mathbb{C}$, where D is a "domain" (the precise meaning of what we mean by a domain will be given in Subsection 1.3.4).

The bulk of the book is then in Chapters 2, 3 and 4, where we construct the following three lanterns to shed light on our central object of study, namely holomorphic functions in a domain:

(1) The Cauchy-Riemann equations,
(2) The Cauchy Integral Theorem,
(3) Taylor series.

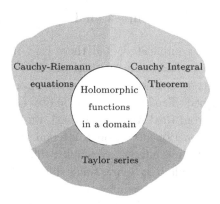

The core content of the book can be summarized in the following Main Theorem[1]:

Theorem 0.2. *Let D be an open path connected set and let $f : D \to \mathbb{C}$. Then the following are equivalent:*

(1) *For all $z \in D$, $f'(z)$ exists.*

(2) *For all $z \in D$ and all $n \geq 0$, $f^{(n)}(z)$ exists.*

(3) $u := \operatorname{Re}(f)$, $v := \operatorname{Im}(f)$ *are continuously differentiable and* $\dfrac{\partial u}{\partial x} = \dfrac{\partial v}{\partial y}$, $\dfrac{\partial u}{\partial y} = -\dfrac{\partial v}{\partial x}$ *in D.*

(4) *For each simply connected subdomain S of D, there exists a holomorphic $F : S \to \mathbb{C}$ such that $F'(z) = f(z)$ for all $z \in S$.*

(5) f *is continuous on D and for all piecewise smooth closed paths γ in each simply connected subdomain of D, we have*

$$\int_{\gamma} f(z)dz = 0.$$

(6) *If $\{z \in \mathbb{C} : |z - z_0| \leq r\} \subset D$, then there is a unique sequence $(c_n)_{n \geq 0}$ in $\mathbb{C}$ such that for all z with $|z - z_0| < r$,*

$$f(z) = \sum_{n=0}^{\infty} c_n(z - z_0)^n.$$

Furthermore, $c_n = \dfrac{1}{2\pi i} \displaystyle\int_{|\zeta - z_0| = r} \dfrac{f(\zeta)}{(\zeta - z_0)^{n+1}} d\zeta$ *and* $c_n = \dfrac{f^{(n)}(z_0)}{n!}$.

---

[1]Don't worry about the unfamiliar terms/notation here: that is what we will learn, besides the proof!

Complex Analysis is not complex analysis!

Indeed, it is not very complicated, and there isn't much analysis. The analysis is "softer" than real analysis: there are fewer deltas and epsilons and difficult estimates, once a few key properties of complex differentiable functions are established. The Main Theorem above tells us that the subject is radically different from Real Analysis. Indeed, we have seen that a real-valued differentiable function on an open interval (a, b) need not have a continuous derivative. In contrast, a complex differentiable function on an open subset of $\mathbb{C}$ is infinitely many times differentiable! This happens because the special geometric meaning of complex multiplication implies that complex differentiable functions behave in a rather controlled manner locally infinitesimally, and aren't allowed to map points willy nilly. This controlled behaviour makes these functions rigid and we will see this in Section 2.3. Nevertheless there are enough of them to make the subject nontrivial and interesting!

The intended audience

These notes constitute a basic course in Complex Analysis, for students who have studied calculus in one and in several variables. The title of the book is meant to indicate that we aim to cover the bare bones of the subject with minimal prerequisites. The notes originated as lecture notes when the second author gave this course for third year students of the BSc programme in Mathematics and/with Economics.

Acknowledgements

Thanks are due to Raymond Mortini, Adam Ostaszewski and Rudolf Rupp for many useful comments. This book relies heavily on some of the sources mentioned in the bibliography. This applies also to the exercises. At some instances we have given detailed references in the section on notes at the end of each chapter, but no claim to originality is made in case there is a missing reference.

Sara Maad Sasane and Amol Sasane,
London and Lund, 2013

Contents

Chapter 1

Complex numbers and their geometry

In this chapter, we set the stage for doing complex analysis. We study three main topics:

(1) We will introduce the set of complex numbers, and their arithmetic, making $\mathbb{C}$ into a field, "extending" the usual field of real numbers.
(2) Points in $\mathbb{C}$ can be depicted in the plane $\mathbb{R}^2$, and we will see that the arithmetic in $\mathbb{C}$ has geometric meaning in the plane. This correspondence between $\mathbb{C}$ and points in the plane also allows one to endow $\mathbb{C}$ with the usual Euclidean topology of the plane.
(3) Finally we will study a fundamental function in complex analysis, namely the exponential function. We also look at some elementary functions related to the exponential function, namely trigonometric functions and the logarithm.

1.1 The field of complex numbers

By definition, a *complex number* is an ordered pair of real numbers. For example,

$$(1,0), \ (0,1), \ (0,0), \ \left(-\frac{3}{4}, \sqrt{2}\right)$$

are all complex numbers. The set $\mathbb{R} \times \mathbb{R}$ of all complex numbers is denoted by $\mathbb{C}$. Thus

$$\mathbb{C} = \{z = (x,y) : x \in \mathbb{R} \text{ and } y \in \mathbb{R}\}.$$

For a complex number $z = (x,y) \in \mathbb{C}$, where $x, y \in \mathbb{R}$, the real number x is called the *real part of z*, and y is called the *imaginary part of z*.

1

We define the operations of *addition* "+" and *multiplication* "·" on $\mathbb{C}$ by:

$$(x_1, y_1) + (x_2, y_2) = (x_1 + x_2, y_1 + y_2),$$
$$(x_1, y_1) \cdot (x_2, y_2) = (x_1 x_2 - y_1 y_2, x_1 y_2 + x_2 y_1),$$

for complex numbers (x_1, y_1), (x_2, y_2). With these operations, $\mathbb{C}$ is a field, that is,

(F1) $(\mathbb{C}, +)$ is an "Abelian group",
(F2) $(\mathbb{C} \setminus \{0\}, \cdot)$ is an Abelian group, and
(F3) the distributive law holds: for $a, b, c \in \mathbb{C}$, $(a + b) \cdot c = a \cdot c + b \cdot c$.

In (F1), Abelian group just means that the operation $+$ on $\mathbb{C}$ is associative, commutative, there exists an "identity element" $(0, 0)$, such that

$$(x, y) + (0, 0) = (x, y) = (0, 0) + (x, y)$$

for all (x, y), and every element (x, y) has an "additive inverse" $(-x, -y)$:

$$(x, y) + (-x, -y) = (0, 0) = (-x, -y) + (x, y).$$

Similarly, in (F2), the multiplicative identity is $(1, 0)$, and the multiplicative inverse of a complex number $(x, y) \in \mathbb{C} \setminus \{(0, 0)\}$ is given by

$$\left(\frac{x}{x^2 + y^2}, \frac{-y}{x^2 + y^2} \right). \tag{1.1}$$

Exercise 1.1. Check that (1.1) is indeed the inverse of $(x, y) \in \mathbb{C} \setminus \{(0, 0)\}$.

We have:

Proposition 1.1. $(\mathbb{C}, +, \cdot)$ *is a field.*

$\mathbb{R}$ **is "contained" in** $\mathbb{C}$. In fact, we can embed $\mathbb{R}$ inside $\mathbb{C}$, and view $\mathbb{R}$ as a "subfield" of $\mathbb{C}$, that is, one can show that the map

$$x \mapsto (x, 0)$$

sending the real number x to the complex number $(x, 0)$ is an injective field homomorphism. This just means that the operations of addition and multiplication are preserved by this map, and distinct real numbers are sent to distinct complex numbers.

$\mathbb{R}$		$\mathbb{C}$
x	$\mapsto$	$(x, 0)$
$x_1 + x_2$	$\mapsto$	$(x_1 + x_2, 0) = (x_1, 0) + (x_2, 0)$
$x_1 \cdot x_2$	$\mapsto$	$(x_1 \cdot x_2, 0) = (x_1, 0) \cdot (x_2, 0)$
1	$\mapsto$	$(1, 0)$
0	$\mapsto$	$(0, 0)$

Thus we can view real numbers as if they are complex numbers via this identification. For example, the real number $\sqrt{2}$ can be viewed as the complex number $(\sqrt{2}, 0)$. If this makes one uneasy, one should note that we have been doing such identifications right from elementary school, where for instance, we identified integers with rational numbers, for example,

$$\mathbb{Z} \ni 3 = \frac{3}{1} \in \mathbb{Q},$$

and we didn't lose sleep over it!

But the advantage of working with $\mathbb{C}$ is that while in $\mathbb{R}$ there was no solution $x \in \mathbb{R}$ to the equation

$$x^2 + 1 = 0,$$

now with complex numbers we have

$$(0,1) \cdot (0,1) + (1,0) = (-1,0) + (1,0) = (0,0).$$

If we give a special symbol, say i, to the number $(0,1)$, then the above says that

$$i^2 + 1 = 0,$$

where we have made the usual identification of the real numbers 1 and 0 with their corresponding complex numbers $(1,0)$ and $(0,0)$.

Henceforth, for the complex number (x, y), where x, y are real, we write $x + yi$, since

$$(x,y) = \underbrace{(x,0)}_{\equiv x} + \underbrace{(y,0)}_{\equiv y} \cdot \underbrace{(0,1)}_{\equiv i} = x + yi.$$

As complex multiplication is commutative, in particular, $yi = iy$, and so we have $x + yi = x + iy$.

Exercise 1.2. Let $\theta \in \left(-\dfrac{\pi}{2}, \dfrac{\pi}{2}\right)$. Express $\dfrac{1 + i\tan\theta}{1 - i\tan\theta}$ in the form $x + yi$, where x, y are real.

Historical development of complex numbers. Contrary to popular belief, historically, it wasn't the need for solving *quadratic* equations, but rather *cubic* equations, that led mathematicians to take complex numbers seriously. The gist of this is the following. Around the sixteenth century, one viewed solving equations like

$$ax^2 + bx + c = 0$$

as the geometric problem of finding the intersection point of the parabola $y = x^2$ with the line $y = -bx - c$. Based on this geometric interpretation, it was easy to dismiss the lack of solvability in reals of a quadratic such as $x^2 + 1 = 0$, since that just reflected the geometric fact that parabola $y = x^2$ did not meet the line $y = -1$. See the picture on the left in Figure 1.1.

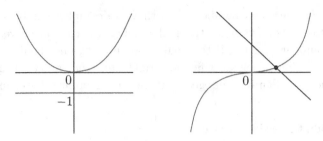

Fig. 1.1 Lack of solvability in reals of $x^2 = -1$ versus the fact that $x^3 = 3px + 2q$ always has a real solution x.

Meanwhile, Cardano (1501-1576) gave a formula for solving the cubic $x^3 = 3px + 2q$, namely,

$$x = \sqrt[3]{q + \sqrt{q^2 - p^3}} + \sqrt[3]{q - \sqrt{q^2 - p^3}}.$$

For example, one can check that for the equation $x^3 = 6x + 6$, with $p = 2$ and $q = 3$, this yields one solution to be $x = \sqrt[3]{4} + \sqrt[3]{2}$. However, note that by the Intermediate Value Theorem, the cubic $y = x^3$ *always* intersects the line $y = 3px + 2q$. See the picture on the right in Figure 1.1. But for an equation like $x^3 = 15x + 4$, that is when $p = 5$ and $q = 2$, we have $q^2 - p^3 = -121 < 0$, and so Cardano's formula fails with real arithmetic, but we do have a real root, namely $x = 4$:

$$4^3 = 64 = 60 + 4 = 15 \cdot 4 + 4.$$

Three decades after the appearance of Cardano's work, Bombelli suggested that maybe with the use of complex arithmetic, Cardano's formula would give the desired real root. So we may ask if

$$x = \sqrt[3]{2 + 11i} + \sqrt[3]{2 - 11i} \overset{?}{=} 4.$$

One can check that $(2+i)^3 = 2+11i$ and $(2-i)^3 = 2-11i$, so that the above does work with these values of the cube root. Thus Bombelli's work established that even for *real* problems, complex arithmetic might be relevant. From then on, complex numbers entered mainstream mathematics.

Exercise 1.3. A field $\mathbb{F}$ is called *ordered* if there is a subset $P \subset \mathbb{F}$, called the *set of positive elements* of $\mathbb{F}$, satisfying the following:

(P1) For all $x, y \in P$, $x + y \in P$.

(P2) For all $x, y \in P$, $x \cdot y \in P$.

(P3) For each $x \in P$, one and only one of the following three cases is true:
$\underline{1}^\circ \quad x = 0. \qquad \underline{2}^\circ \quad x \in P. \qquad \underline{3}^\circ \quad -x \in P.$

For example, the field of real numbers $\mathbb{R}$ is ordered, since $P := (0, \infty)$ is a set of positive elements of $\mathbb{R}$. (Once one has an ordered set of elements in a field, one can compare the elements of $\mathbb{F}$ by defining a relation $>_P$ in $\mathbb{F}$ by setting $y >_P x$ for $x, y \in \mathbb{F}$ if $y - x \in P$.) Show that $\mathbb{C}$ is not an ordered field.
Hint: Consider $x := i$, and first look at $x \cdot x$.

1.2 Geometric representation of complex numbers

Since $\mathbb{C} = \mathbb{R}^2$, we can identify complex numbers with points in the plane. See Figure 1.2.

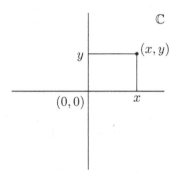

Fig. 1.2 The complex number $x + iy$ in the complex plane.

The complex plane is sometimes called the *Argand*[1] *plane*.

Exercise 1.4. Locate the following points in the complex plane: 0, 1, $-\dfrac{3}{2}$, i, $-\sqrt{2}i$, $\cos\dfrac{\pi}{3} + i\sin\dfrac{\pi}{3}$.

So we can identify $\mathbb{C}$ as a *set* with the points in the plane $\mathbb{R}^2$. Do the field operations in $\mathbb{C}$ have some geometric meaning in the plane? We see below that this is indeed the case: addition in $\mathbb{C}$ is vector addition in the plane, and multiplication in $\mathbb{C}$ has a special geometric meaning in the plane, which is explained below as well.

Geometric meaning of complex addition. With this identification of complex numbers with points in the plane, it is clear that complex addition is just addition of vectors in $\mathbb{R}^2$. By addition of vectors, we mean the

---

[1]It is named after Jean-Robert Argand (1768-1822), although it was used earlier by Caspar Wessel (1745-1818).

usual way of combining two vectors, that is, by completing the parallelogram formed by the line segments joining $(0,0)$ to each of the two complex numbers as sides, and then taking the endpoint of the diagonal from $(0,0)$ as the sum of the two given complex numbers. See Figure 1.3.

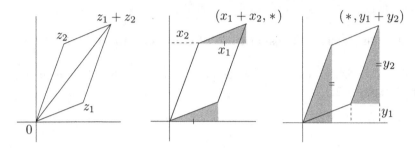

Fig. 1.3 Addition of complex numbers is vector addition in $\mathbb{R}^2$.

Indeed, the middle picture shows that addition of z_1 and z_2 as vectors in the plane yields the correct x-coordinate of their sum as complex numbers by looking at the two congruent blue triangles. Similarly, by the rightmost picture shows that the y-coordinate is also correct.

Geometric meaning of complex multiplication. We will now see the special geometric meaning of complex multiplication. In order to do this, it will be convenient to use polar coordinates. Thus, let the point $(x,y) \in \mathbb{R}^2$ have polar coordinates $r \geq 0$ and $\theta \in (-\pi, \pi]$. This means that the distance of the point (x,y) to $(0,0)$ is r (≥ 0), and the angle made by the ray from $(0,0)$ to (x,y) makes an angle of θ with the positive real axis (the x-axis). (If (x,y) is itself $(0,0)$, we set $\theta = 0$.)

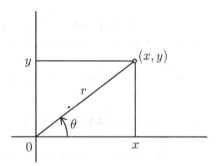

Fig. 1.4 Polar coordinates (r, θ) of $(x,y) \in \mathbb{R}^2$.

Then from the right-angled triangle shown in Figure 1.4, we have

$$x = r \cos \theta,$$

$$y = r \sin \theta.$$

Thus we can express the complex number in terms of the polar coordinates (r, θ):

$$x + yi = r \cos \theta + (r \sin \theta)i = r(\cos \theta + i \sin \theta).$$

Now we give the geometric interpretation of complex multiplication. For two complex numbers expressed in polar coordinates as

$$z_1 = r_1(\cos \theta_1 + i \sin \theta_1),$$

$$z_2 = r_2(\cos \theta_2 + i \sin \theta_2),$$

we have that

$$
\begin{aligned}
z_1 \cdot z_2 &= r_1(\cos \theta_1 + i \sin \theta_1) \cdot r_2(\cos \theta_2 + i \sin \theta_2) \\
&= r_1 r_2(\cos \theta_1 \cos \theta_2 - \sin \theta_1 \sin \theta_2 + i(\cos \theta_1 \sin \theta_2 + \cos \theta_2 \sin \theta_1)) \\
&= r_1 r_2(\cos(\theta_1 + \theta_2) + i \sin(\theta_1 + \theta_2)),
\end{aligned}
$$

using the trigonometric identities for angle addition. Thus $z_1 \cdot z_2$ has the polar coordinates $(r_1 r_2, \theta_1 + \theta_2)$. In other words, the angles z_1 and z_2 make with the positive real axis are *added* in order to get the angle $z_1 \cdot z_2$ makes with the positive real axis, and the distances to the origin are *multiplied* to get the distance $z_1 \cdot z_2$ has to the origin. See Figure 1.5.

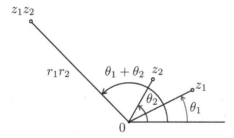

Fig. 1.5 Geometric meaning of complex multiplication: angles get added, distances to the origin get multiplied.

As a special case, consider multiplication by $\cos \alpha + i \sin \alpha$, which is at a distance of 1 from the origin. Then from the above, we see that if $z \in \mathbb{C}$, then $z \cdot (\cos \alpha + i \sin \alpha)$ is obtained by rotating the line joining 0 to z anticlockwise through an angle of α. In particular, multiplying z by

$$i = 0 + i \cdot 1 = \cos \frac{\pi}{2} + i \sin \frac{\pi}{2}$$

produces a counterclockwise rotation of $90°$.

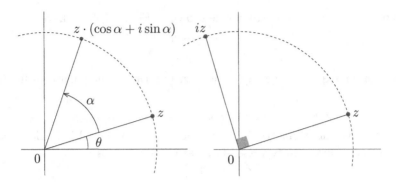

Fig. 1.6   Multiplication by $\cos\alpha + i\sin\alpha$ produces an anticlockwise rotation through α.

De Moivre's formula and nth roots. We have for all $n \in \mathbb{N}$

$$(\cos\theta + i\sin\theta)^n = \cos(n\theta) + i\sin(n\theta).$$

This is called *de Moivre's formula*.

Exercise 1.5. Recover the trigonometric equality $\cos(3\theta) = 4(\cos\theta)^3 - 3\cos\theta$ using de Moivre's formula.

Exercise 1.6. Express $(1+i)^{10}$ in the form $x + iy$ with real x, y without expanding!

Exercise 1.7. By considering $(2 + i)(3 + i)$, show that $\dfrac{\pi}{4} = \tan^{-1}\dfrac{1}{2} + \tan^{-1}\dfrac{1}{3}$.

Exercise 1.8. *Gaussian integers* are complex numbers of the form $m + in$, where m, n are integers. Thus they are integral lattice points in the complex plane. Show that it is impossible to draw an equilateral triangle such that all vertices are Gaussian integers.
Hint: Rotation of one of the sides should give the other. Recall that $\sqrt{3} \notin \mathbb{Q}$.

De Moivre's formula gives an easy way of finding the *nth roots of a complex number z*, that is, complex numbers w that satisfy $w^n = z$. Indeed, we first write $z = r(\cos\theta + i\sin\theta)$ for some $r \geq 0$ and $\theta \in [0, 2\pi)$. Now if $w^n = z$, where $w = \rho(\cos\alpha + i\sin\alpha)$, then

$$w^n = \rho^n(\cos(n\alpha) + i\sin(n\alpha)) = r(\cos\theta + i\sin\theta) = z,$$

and so by equating the distance to the origin on both sides, we obtain $\rho^n = r$. Hence $\rho = \sqrt[n]{r}$, as both ρ and r are nonnegative. On the other hand, the angle that w^n makes with the positive real axis is $n\alpha$, which is in the set $\{\cdots,\ \theta - 4\pi,\ \theta - 2\pi,\ \theta,\ \theta + 2\pi,\ \theta + 4\pi,\ \theta + 6\pi,\ \cdots\}$, because the angle made by a nonzero z with the positive real axis is unique only up to integral multiples of 2π, that is, instead of θ, we could just as well have used $\theta + 2\pi k$ for any integer k. See Figure 1.7.

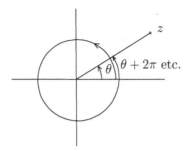

Fig. 1.7 The angle made by z with the positive real axis.

Thus we get that $\alpha \in \left\{ \dfrac{\theta}{n} + \dfrac{2\pi}{n} k : k \in \mathbb{Z} \right\}$, and this gives distinct w for

$$\alpha \in \left\{ \frac{\theta}{n}, \ \frac{\theta}{n} + \frac{2\pi}{n}, \ \frac{\theta}{n} + 2 \cdot \frac{2\pi}{n}, \ \ldots, \ \frac{\theta}{n} + (n-1) \cdot \frac{2\pi}{n} \right\}.$$

In particular, if $z = 1$, we get the nth roots of unity, which are located at the vertices of an n-sided regular polygon inscribed in a circle. See Figure 1.8.

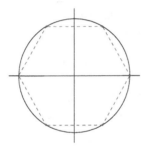

Fig. 1.8 The six 6th roots of unity.

Exercise 1.9. Find all complex numbers w such that $w^4 = -1$. Depict these in the complex plane.

Exercise 1.10. Find all complex numbers z that satisfy $z^6 - z^3 - 2 = 0$.

Exercise 1.11. If a, b, c are *real* numbers such that $a^2 + b^2 + c^2 = ab + bc + ca$, then they must be equal. Indeed, doubling both sides and rearranging gives $(a - b)^2 + (b - c)^2 + (c - a)^2 = 0$, and since each summand is nonnegative, it must be the case that each is 0. On the other hand, now show that if a, b, c are *complex* numbers such that $a^2 + b^2 + c^2 = ab + bc + ca$, then they must lie on the vertices of an equilateral triangle in the complex plane. Explain the real case result in light of this fact.
Hint: Calculate $((b - a)\omega + (b - c)) \cdot ((b - a)\omega^2 + (b - c))$, where ω is a nonreal cube root of unity.

Exercise 1.12. The Binomial Theorem says that if a, b are real numbers and $n \in \mathbb{N}$, then

$$(a+b)^n = \sum_{k=0}^{n} \binom{n}{k} a^k b^{n-k}, \quad \text{where} \quad \binom{n}{k} := \frac{n!}{k!(n-k)!}, \quad k = 0, 1, 2, \ldots, n,$$

are the binomial coefficients. The algebraic reasoning leading to this is equally valid if a, b are complex numbers. Show that

$$\binom{3n}{0} + \binom{3n}{3} + \binom{3n}{6} + \cdots + \binom{3n}{3n} = \frac{2^{3n} + 2 \cdot (-1)^n}{3}.$$

Hint: Find $(1+1)^{3n} + (1+\omega)^{3n} + (1+\omega^2)^{3n}$, where ω denotes a nonreal cube root of unity.

Exercise 1.13. Show, using the geometry of complex numbers, that the line segments joining the centers of opposite external squares described on sides of an arbitrary convex quadrilateral are perpendicular and have equal lengths.

Absolute value and complex conjugate. The *absolute value* $|z|$ of the complex number $z = x + iy$, where $x, y \in \mathbb{R}$, is defined by

$$|z| = \sqrt{x^2 + y^2}.$$

Note that by Pythagoras' Theorem, this is the distance of the complex number z to 0 in the complex plane. See the picture on the left of Figure 1.9. By expressing $z_1, z_2 \in \mathbb{C}$ in terms of polar coordinates, or by a direct calculation, it is clear that $|z_1 z_2| = |z_1| \cdot |z_2|$.

Exercise 1.14. Verify the property $|z_1 z_2| = |z_1| \cdot |z_2|$ by expressing z_1, z_2 using Cartesian coordinates.

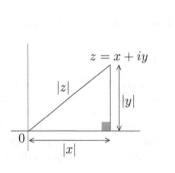 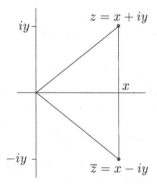

Fig. 1.9   The absolute value of z is the distance of z to the origin, and the complex conjugate is obtained by reflecting z in the real axis.

The *complex conjugate* $\bar{z}$ of $z = x + iy$ where $x, y \in \mathbb{R}$, is defined by

$$\bar{z} = x - iy.$$

In the complex plane, $\bar{z}$ is obtained by reflecting the point corresponding to z in the real axis. See the picture on the right of Figure 1.9. From this geometric interpretation, convince yourself that for all $z_1, z_2 \in \mathbb{C}$,

$$\overline{z_1 + z_2} = \overline{z_1} + \overline{z_2} \quad \text{and} \quad \overline{z_1 \cdot z_2} = \overline{z_1} \cdot \overline{z_2}.$$

The following properties are easy to check:

$$\bar{\bar{z}} = z, \quad z\bar{z} = |z|^2, \quad \mathrm{Re}(z) = \frac{z + \bar{z}}{2}, \quad \mathrm{Im}(z) = \frac{z - \bar{z}}{2i}.$$

Exercise 1.15. Verify that the four equalities above hold.

Exercise 1.16. Prove that for all $z \in \mathbb{C}$, $|z| = |\bar{z}|$, $|\mathrm{Re}(z)| \leq |z|$ and $|\mathrm{Im}(z)| \leq |z|$. Give geometric interpretations of each.

Exercise 1.17. If $a, z \in \mathbb{C}$ satisfy $|a| < 1$ and $|z| \leq 1$, then prove that $\left| \dfrac{z - a}{1 - \bar{a}z} \right| \leq 1$.

Exercise 1.18. Consider the polynomial p given by $p(z) = c_0 + c_1 z + \cdots + c_d z^d$, where $c_0, c_1, \ldots, c_d \in \mathbb{R}$ and $c_d \neq 0$. Show that if $w \in \mathbb{C}$ is such that $p(w) = 0$, then also $p(\bar{w}) = 0$.

Exercise 1.19. Show that the area of the triangle formed by $0, a, b \in \mathbb{C}$ is $\left| \dfrac{\mathrm{Im}(a\bar{b})}{2} \right|$.

Exercise 1.20. Prove for any complex z_1, z_2, z_3 that $i \det \begin{bmatrix} 1 & z_1 & \overline{z_1} \\ 1 & z_2 & \overline{z_2} \\ 1 & z_3 & \overline{z_3} \end{bmatrix}$ is real.

Exercise 1.21. Show that for any two complex numbers z_1, z_2, there holds that $|z_1 + z_2|^2 + |z_1 - z_2|^2 = 2(|z_1|^2 + |z_2|^2)$. What is the geometric interpretation of this equality?

1.3   Topology of $\mathbb{C}$

The concepts in ordinary calculus in the setting of $\mathbb{R}$, like convergence of sequences, or continuity and differentiability of functions, all rely on the notion of closeness of points in $\mathbb{R}$. For example, when we talk about the convergence of a real sequence $(a_n)_{n \in \mathbb{N}}$ to its limit $L \in \mathbb{R}$, we mean that given any positive ϵ, there is a large enough index N such that beyond that index, the corresponding terms a_n all have a *distance* to L which is at most

ϵ. This "distance of a_n to L" is taken as $|a_n - L|$, and this is the length of the line segment joining the numbers a_n and L on the real number line.

Now in order to do calculus with *complex* numbers, we need a notion of distance $d(z_1, z_2)$ between for pairs of complex numbers (z_1, z_2), and the first order of business is to explain what this notion is.

1.3.1 *Metric on* $\mathbb{C}$

Since $\mathbb{C}$ is just $\mathbb{R}^2$, we use the usual Euclidean distance in $\mathbb{R}^2$ as the metric in $\mathbb{C}$. Thus, for complex numbers $z_1 = x_1 + iy_1$ and $z_2 = x_2 + iy_2$, we have

$$d(z_1, z_2) = \sqrt{(x_1 - x_2)^2 + (y_1 - y_2)^2} = |z_1 - z_2|.$$

By Pythagoras's Theorem, this is the length of the line segment joining the points $(x_1, y_1), (x_2, y_2)$ in $\mathbb{R}^2$; see Figure 1.10.

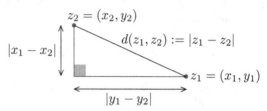

Fig. 1.10   The distance between z_1 and z_2 is the length of the segment joining z_1 to z_2.

Using the geometric meaning of addition of complex numbers, and the well known result from Euclidean geometry that the sum of the lengths of any two sides of a triangle is at least as big as the length of the third side, we obtain the following *triangle inequality* for the absolute value:

$$|z_1 + z_2| \le |z_1| + |z_2| \text{ for } z_1, z_2 \in \mathbb{C}.$$

See Figure 1.11. The triangle inequality above can also be verified analytically by using the Cauchy-Schwarz inequality for real numbers x_1, x_2, y_1, y_2: $(x_1^2 + y_1^2)(x_2^2 + y_2^2) \ge (x_1 x_2 + y_1 y_2)^2$.

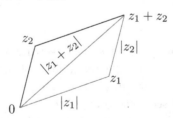

Fig. 1.11   Triangle inequality.

Exercise 1.22. Show that for all $z_1, z_2 \in \mathbb{C}$, $|z_1 - z_2| \geq ||z_1| - |z_2||$.

Exercise 1.23. Sketch the following sets in the complex plane:

(1) $\{z \in \mathbb{C} : |z - (1 - i)| = 2\}$.

(2) $\{z \in \mathbb{C} : |z - (1 - i)| < 2\}$.

(3) $\{z \in \mathbb{C} : 1 < |z - (1 - i)| < 2\}$.

(4) $\{z \in \mathbb{C} : \text{Re}(z - (1 - i)) = 3\}$.

(5) $\{z \in \mathbb{C} : |\text{Im}(z - (1 - i))| < 3\}$.

(6) $\{z \in \mathbb{C} : |z - (1 - i)| = |z - (1 + i)|\}$.

(7) $\{z \in \mathbb{C} : |z - (1 - i)| + |z - (1 + i)| = 2\}$.

(8) $\{z \in \mathbb{C} : |z - (1 - i)| + |z - (1 + i)| < 3\}$.

1.3.2 *Open discs, open sets, closed sets and compact sets*

In order to talk about sets of points near a given point, it will be convenient to introduce the following definitions.

An *open ball/disc $D(z_0, r)$ with center z_0 and radius $r > 0$* is defined by $D(z_0, r) := \{z \in \mathbb{C} : |z - z_0| < r\}$.

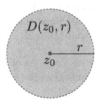

A subset U of $\mathbb{C}$ is called *open* if for every $z \in U$, there exists an $r_z > 0$ such that $D(z, r_z) \subset U$. In other words, no matter what point we choose in U, there is always some "room" around that point comprising points only from U. For example, it can be checked that an open disc $D(z_0, r)$ is an open set. So using the adjective *open* in the name for $D(z_0, r)$ makes sense. Here are some more examples of open sets. The annulus $\mathbb{A}_r := \{z \in \mathbb{C} : r < |z| < 1\}$ and the right half-plane $\mathbb{H} := \{z \in \mathbb{C} : \text{Re}(z) > 0\}$ are open sets.

It is also convenient to give a special name to sets whose complement is open, and these are called *closed sets*. One can also give a characterization of closed sets in terms of sequential convergence: A set $F \subset \mathbb{C}$ is closed if and only if for every sequence $(z_n)_{n \in \mathbb{N}}$ in F which is convergent in $\mathbb{C}$ to L, there holds that the limit $L \in F$.

A subset S of $\mathbb{C}$ is called *bounded* if there exists a $M > 0$ such that for all $z \in S$, $|z| \leq M$. Thus S is contained in a big enough disc in the complex plane. A subset $K \subset \mathbb{C}$ is called *compact* if it is both closed and bounded. We will often use the known fact from real analysis that a real valued continuous function on a compact set possesses a maximizer and a minimizer.

1.3.3   *Convergence and continuity*

We can also talk about convergent sequences in $\mathbb{C}$.

A sequence $(z_n)_{n \in \mathbb{N}}$ is said to be *convergent with limit* L if for every $\epsilon > 0$, there exists an index $N \in \mathbb{N}$ such that for every $n > N$, there holds that $|z_n - L| < \epsilon$. It follows from the triangle inequality that for a convergent sequence the limit is unique, and we write

$$\lim_{n \to \infty} z_n = L.$$

Example 1.1. Let z be a complex number with $|z| < 1$. Then the sequence $(z^n)_{n \in \mathbb{N}}$ converges to 0. Indeed, $|z^n - 0| = |z^n| = |z|^n = ||z|^n - 0|$, but as $|z| < 1$, we know that $|z|^n \to 0$ as $n \to \infty$. $\Diamond$

Exercise 1.24. Consider the polynomial p given by $p(z) = c_0 + c_1 z + \cdots + c_d z^d$, where $c_0, c_1, \ldots, c_d \in \mathbb{C}$ and $c_d \neq 0$. Show that there exist $M, R > 0$ such that $|p(z)| \geq M|z|^d$ for all $z \in \mathbb{C}$ such that $|z| > R$.

Exercise 1.25. Show that a sequence $(z_n)_{n \in \mathbb{N}}$ of complex numbers is convergent to L if and only if the two real sequences $(\mathrm{Re}(z_n))_{n \in \mathbb{N}}$ and $(\mathrm{Im}(z_n))_{n \in \mathbb{N}}$ are convergent respectively to $\mathrm{Re}(L)$ and $\mathrm{Im}(L)$.

Exercise 1.26. Show that a sequence $(z_n)_{n \in \mathbb{N}}$ of complex numbers is convergent to L if and only if $(\overline{z_n})_{n \in \mathbb{N}}$ converges to $\overline{L}$.

Exercise 1.27. Prove that $\mathbb{C}$ is complete, that is, every Cauchy sequence in $\mathbb{C}$ converges in $\mathbb{C}$. (A sequence $(z_n)_{n \in \mathbb{N}}$ is called a *Cauchy sequence* if for every $\epsilon > 0$, there is an index $N \in \mathbb{N}$ such that for all indices $m, n > N$, there holds that $|z_n - z_m| < \epsilon$.)

Let S be a subset of $\mathbb{C}$, $z_0 \in S$ and $f : S \to \mathbb{C}$. Then f is said to be *continuous at* z_0 if for every $\epsilon > 0$, there exists a $\delta > 0$ such that whenever $z \in S$ satisfies $|z - z_0| < \delta$, there holds that $|f(z) - f(z_0)| < \epsilon$. f is said to be *continuous* if for every $z \in S$, f is continuous at z.

One can also give a characterization of continuity at a point in terms of convergent sequences: $f : S \to \mathbb{C}$ is continuous at $z_0 \in S$ if and only if for every sequence $(z_n)_{n\in\mathbb{N}}$ in S convergent to z_0, there holds that the sequence $(f(z_n))_{n\in\mathbb{N}}$ is convergent to $f(z_0)$.

Example 1.2. Complex conjugation is continuous, that is, $z \mapsto \overline{z} : \mathbb{C} \to \mathbb{C}$ is continuous. Indeed, we have $|\overline{z} - \overline{z_0}| = |\overline{z - z_0}| = |z - z_0|$ for all $z, z_0 \in \mathbb{C}$. This shows that complex conjugation is continuous at each $z_0 \in \mathbb{C}$, and so it is a continuous mapping. This is geometrically obvious, since complex conjugation is just reflection in the real axis, and so the image stays close to the reflected point if we are close to the point!

Since $\overline{(\overline{z})} = z$ for all $z \in \mathbb{C}$, complex conjugation is its own inverse. So complex conjugation is invertible with a continuous inverse. Thus complex conjugation gives a homeomorphism (that is, a continuous bijective mapping with a continuous inverse) from $\mathbb{C}$ to $\mathbb{C}$. $\Diamond$

Exercise 1.28. Prove that the map $z \mapsto \mathrm{Re}(z) : \mathbb{C} \to \mathbb{R}$ is continuous.

1.3.4   *Domains*

In the sequel, the notion of a *path-connected open set* will play an important role. By this we mean that we will prove results about our central object of study, namely functions $f : D \to \mathbb{C}$ that are complex differentiable at every point of the set D ($\subset \mathbb{C}$), and it will turn out that for the validity of many of these theorems, we will need D to be a "nice" subset of $\mathbb{C}$, and not just any old subset of $\mathbb{C}$. Sets which satisfy this "niceness" assumption, stipulated precisely below, will be what we call a domain.

We will call an open path-connected subset of $\mathbb{C}$ a *domain*. We already know what "open" means. Now let us explain what we mean by "path-connectedness".

Definition 1.1.

(1) A *path* (or *curve*) in $\mathbb{C}$ is a continuous function $\gamma : [a, b] \to \mathbb{C}$.

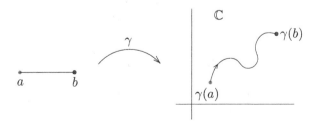

(2) A *stepwise path* is a path $\gamma : [a, b] \to \mathbb{C}$ such that there are points $t_0 = a < t_1 < \cdots < t_n < t_{n+1} = b$ such that for each $k = 0, 1, \ldots, n$, the restriction $\gamma : [t_k, t_{k+1}] \to \mathbb{C}$ is a path with either a constant real part or constant imaginary part.

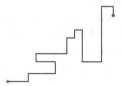

(3) An open set $U \subset \mathbb{C}$ is called *path-connected* if for every $z_1, z_2 \in U$, there is a stepwise path $\gamma : [a, b] \to \mathbb{C}$ such that $\gamma(a) = z_1$, $\gamma(b) = z_2$, and for all $t \in [a, b]$, $\gamma(t) \in U$.

Actually in the above definition of path-connected open sets, the restriction that the paths being *stepwise* paths can be relaxed, that is, if we look at those open sets in which any two points can be joined merely by a path, then this class of open sets coincides with our path-connected sets. But this is an unnecessary diversion for us. So we just live with the definition we have given instead.

Example 1.3.

(1) The open unit disc $\mathbb{D} := \{z \in \mathbb{C} : |z| < 1\}$ is a domain.
(2) For $r \in (0, 1)$, the annulus $\mathbb{A}_r := \{z \in \mathbb{C} : r < |z| < 1\}$ is a domain.
(3) The right half-plane $\mathbb{H} := \{z \in \mathbb{C} : \mathrm{Re}(z) > 0\}$ is a domain.

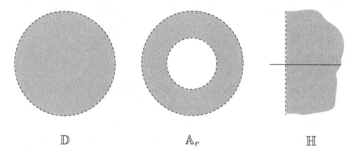

$\mathbb{D}$ $\qquad\qquad\qquad$ $\mathbb{A}_r$ $\qquad\qquad\qquad$ $\mathbb{H}$

Fig. 1.12   The domains $\mathbb{D}$, $\mathbb{A}_r$ and $\mathbb{H}$.

On the other hand, the set $S := \{z \in \mathbb{C} : |z| \neq 1\} =: \mathbb{C} \setminus \mathbb{T}$ is not a domain, since although it is open, it is not path-connected. Indeed, there is no path joining, say 0 and 2: for if there were one such path γ, then by the Intermediate Value Theorem applied to the map $t \mapsto |\gamma(t)| : [a, b] \to \mathbb{R}$, we see that since $|\gamma(a)| = 0 < 1 < 2 = |\gamma(b)|$, there must be a $t_* \in [a, b]$ such that $|\gamma(t_*)| = 1$, but then $\gamma(t_*) \notin S$. $\qquad\diamond$

Exercise 1.29. Show that the set $\{z \in \mathbb{C} : \operatorname{Re}(z) \cdot \operatorname{Im}(z) > 1\}$ is open, but not a domain.

Exercise 1.30. Let D be a domain. Set $D^* := \{z \in \mathbb{C} : \overline{z} \in D\}$. Show that D^* is also a domain.

1.4 The exponential function and kith

In this last section, we discuss some basic complex functions:

the exponential function $z \mapsto \exp z$,
the trigonometric functions $z \mapsto \sin z$, $\cos z$,
and the logarithm $z \mapsto \operatorname{Log} z$.

They will serve as counterparts to the familiar functions from calculus, to which they reduce when restricted to the real axis. In other words, when we restrict our functions to the argument $z = x \in \mathbb{R}$, then we get the usual real-valued functions

$$x \mapsto e^x,$$
$$x \mapsto \sin x, \cos x,$$
$$x \mapsto \log x.$$

So our definitions provide *extensions* of the usual real-valued counterparts; see Figure 1.13.

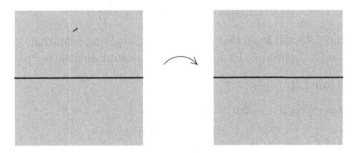

Fig. 1.13   The real valued functions map points on the real line to the real line, but our definitions will give extensions of these to the shaded region, the complex plane.

We will see that these extensions have new and interesting properties in the complex domain that are not possessed by them when the argument is only allowed to be real. Also they will serve as important examples of complex differentiable functions: we will see later on that the exponential and trigonometric functions are complex differentiable everywhere in the complex plane, and the logarithm function is complex differentiable where it happens to be continuous.

Let's begin with the exponential function.

1.4.1   *The exponential* exp z

Definition 1.2 (The complex exponential). *For $z = x+iy \in \mathbb{C}$, where x, y are real, we define the* complex exponential, *denoted by* $\exp z$, *as follows:*

$$\exp z = e^x(\cos y + i \sin y).$$

First we note that when $y = 0$, the right hand side is simply e^x. So our definition extends the usual exponential function ($\mathbb{R} \ni$) $x \mapsto e^x$ ($\in \mathbb{R}$). But this definition does appears to be mysterious. After all, $z \mapsto e^{\mathrm{Re}(z)}$ also gives an extension of the usual real exponential function. So why not use this simple definition instead? We define our exp function in the way we do, because we will see later on that this is the *unique* extension of the real exponential to the whole complex plane having the property that the extension is complex differentiable everywhere; see Example 4.8 on page 128. In fact, just like the real counterpart, where we have that

$$\frac{d}{dx}e^x = e^x \text{ for all } x \in \mathbb{R},$$

we will later see that

$$\frac{d}{dz}\exp z = \exp z \text{ for all } z \in \mathbb{C}.$$

So eventually we will learn that our mysterious looking definition is actually quite natural! Right now, let us check the following elementary properties:

Proposition 1.2.

(1) $\exp 0 = e^0(\cos 0 + i \sin 0) = 1 \cdot (1 + i0) = 1$.
(2) *For $z_1, z_2 \in \mathbb{C}$,* $\exp(z_1 + z_2) = (\exp z_1)(\exp z_2)$.
(3) *For $z \in \mathbb{C}$,* $\exp z \neq 0$, *and* $(\exp z)^{-1} = \exp(-z)$.
(4) *For $z \in \mathbb{C}$,* $\exp(z + 2\pi i) = \exp z$.
(5) *For $z \in \mathbb{C}$,* $|\exp z| = e^{\mathrm{Re}(z)}$.

Proof. (2) If $z_1 = x_1 + iy_1$ and $z_2 = x_2 + iy_2$, then

$\exp(z_1 + z_2)$
$$= e^{(x_1+x_2)+i(y_1+y_2)} = e^{x_1+x_2}(\cos(y_1 + y_2) + i\sin(y_1 + y_2))$$
$$= e^{x_1}e^{x_2}(\cos y_1 \cos y_2 - \sin y_1 \sin y_2 + i(\sin y_1 \cos y_2 + \cos y_1 \sin y_2))$$
$$= e^{x_1}(\cos y_1 + i\sin y_1)e^{x_2}(\cos y_2 + i\sin y_2) = (\exp z_1)(\exp z_2).$$

(3) From the previous part, we see that

$$1 = \exp 0 = \exp(z - z) = (\exp z)(\exp(-z)),$$

showing that $\exp z \neq 0$ and $(\exp z)^{-1} = \exp(-z)$. Thus exp maps $\mathbb{C}$ to the "punctured" plane $\mathbb{C} \setminus \{0\}$.

(4) We have

$$\exp(z + 2\pi i) = (\exp z)(\exp(2\pi i)) = (\exp z) \cdot e^0(\cos(2\pi) + i\sin(2\pi))$$
$$= (\exp z) \cdot 1 \cdot (1 + i \cdot 0) = \exp z.$$

This shows that exp is "periodic in the y-direction" in the complex plane, with a period of 2π; see Figure 1.14.

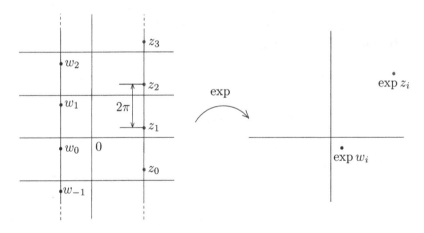

Fig. 1.14 2π-periodicity of exp in the y-direction.

This phenomenon is not present in the x-direction, where just like in the real setting, the function $x \mapsto \exp(x+iy_0)$ (with $y_0 \in \mathbb{R}$ fixed) is one-to-one. See Figure 1.15.

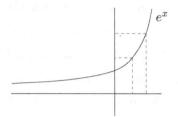

Fig. 1.15 $x \mapsto e^x$ is one-to-one.

(5) For $x, y \in \mathbb{R}$, $|e^x \cos y + i e^x \sin y| = \sqrt{e^{2x}((\cos y)^2 + (\sin y)^2)} = e^x$. So $|\exp(x+iy)| = e^x$. This implies that exp maps vertical lines in the complex plane (that is all points having a common real part) into circles (that is all points having the same absolute value, in other words same distance to the origin). □

Proposition 1.2.(3) above shows that the map exp is *not* one-to-one, but rather, it is periodic with period $2\pi i$. Figure 1.16 shows the effect of the mapping $z \mapsto \exp z$ on horizontal (fixed imaginary part y) and vertical lines (fixed real part x). This picture is arrived at by putting together the observations displayed in Figure 1.17.

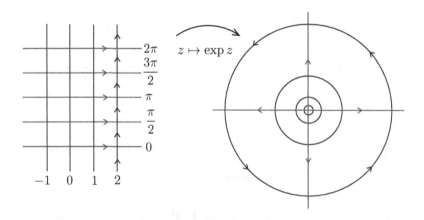

Fig. 1.16   The image of horizontal and vertical lines under the exponential map.

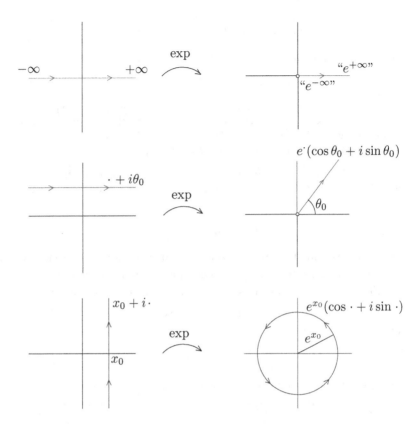

Fig. 1.17   The image of horizontal and vertical lines under the exponential map.

In Figure 1.16, we note that exp preserves the angle between the curves we have considered in its domain. Namely, the horizontal and vertical lines, which are mutually perpendicular, are mapped to circles and radial rays, which are also mutually perpendicular. We will see later on that this is no coincidence, and this property of "conformality", that is, of the preservation of angles between curves in the domain together with their "orientation" is something which is possessed by all complex differentiable functions in domains.

Euler's formula. Note that for $z = iy$, where y is real, we have

$$\exp(iy) = \cos y + i \sin y.$$

This is the so-called *Euler's formula*. Hence the polar form of a complex number can now be rewritten as $z = r(\cos \theta + i \sin \theta) = r \exp(i\theta)$.

Exercise 1.31. Compute $\exp z$ for the following values of z: $i\dfrac{9\pi}{2}$, $3 + \pi i$.

Exercise 1.32. Find all $z \in \mathbb{C}$ that satisfy $\exp z = \pi i$.

Exercise 1.33. Plot the curve $t \mapsto \exp(it) : [0, 2\pi] \to \mathbb{C}$.

Exercise 1.34. Describe the image of the line $y = x$ under the exponential map $z = x + iy \mapsto \exp z$. Proceed as follows: Start with the parametric form $x = t$, $y = t$, and get an expression for the image curve in parametric form. Plot this curve, explaining what happens when t increases, and when $t \to \pm\infty$.

Exercise 1.35. Find the modulus and the real and imaginary parts of $\exp(z^2)$ and of $\exp(1/z)$ in terms of the real and imaginary parts x, y of $z = x + iy$.

1.4.2 *Trigonometric functions*

Just as we extended the real exponential function, we now extend the familiar real trigonometric functions to complex trigonometric functions. From the Euler formula we established earlier, we have for real x that

$$\exp(ix) = \cos x + i \sin x \quad \text{and} \quad \exp(-ix) = \cos x - i \sin x,$$

which gives

$$\cos x = \frac{\exp(ix) + \exp(-ix)}{2} \quad \text{and} \quad \sin x = \frac{\exp(ix) - \exp(-ix)}{2i}.$$

This prompts the following definitions: for $z \in \mathbb{C}$, we *define*

$$\cos z = \frac{\exp(iz) + \exp(-iz)}{2} \quad \text{and} \quad \sin z = \frac{\exp(iz) - \exp(-iz)}{2i}.$$

Clearly these definitions give extensions of the usual real trigonometric functions because when we put $z = x$, we get $\cos z = \cos x$ and $\sin z = \sin x$, as we had seen above from Euler's formula.

Several trigonometric identities continue to hold in the complex setting. For instance, $\cos(z_1 + z_2) = (\cos z_1)(\cos z_2) - (\sin z_1)(\sin z_2)$. Indeed,

$$(\cos z_1)(\cos z_2) - (\sin z_1)(\sin z_2)$$

$$= \left(\frac{\exp(iz_1) + \exp(-iz_1)}{2} \right) \left(\frac{\exp(iz_2) + \exp(-iz_2)}{2} \right)$$

$$- \left(\frac{\exp(iz_1) - \exp(-iz_2)}{2i} \right) \left(\frac{\exp(iz_2) - \exp(-iz_2)}{2i} \right)$$

$$= \frac{2\exp(i(z_1 + z_2)) + 2\exp(-i(z_1 + z_2))}{4} = \cos(z_1 + z_2).$$

Exercise 1.36. Show that $\sin(z_1 + z_2) = (\sin z_1)(\cos z_2) + (\cos z_1)(\sin z_2)$ for all $z_1, z_2 \in \mathbb{C}$.

Also, $(\sin z)^2 + (\cos z)^2 = 1$, since

$$(\sin z)^2 + (\cos z)^2 = \left(\frac{\exp(iz) - \exp(-iz)}{2i}\right)^2 + \left(\frac{\exp(iz) + \exp(-iz)}{2}\right)^2$$

$$= \frac{\exp(2iz) - 2 + \exp(-2iz)}{-4} + \frac{\exp(2iz) + 2 + \exp(-2iz)}{4}$$

$$= 1.$$

However, as opposed to the real trigonometric functions which satisfy $|\sin x| \le 1$ and $|\cos x| \le 1$ for real x, $z \mapsto \sin z$ and $z \mapsto \cos z$ are *not* bounded. Indeed, for $z = iy$, where y is real, we have

$$\cos(iy) = \frac{\exp(i(iy)) + \exp(-i(iy))}{2} = \frac{\exp(-y) + \exp(y)}{2} = \frac{e^{-y} + e^y}{2},$$

and so $\cos(iy) \to +\infty$ as $y \to \pm\infty$. Similarly, since

$$\sin(iy) = \frac{e^{-y} - e^y}{2i}$$

we also have that $|\sin(iy)| \to \infty$ as $y \to \pm\infty$.

We will see later that $z \mapsto \cos z$ and $z \mapsto \sin z$ are complex differentiable everywhere in the complex plane.

Exercise 1.37. Show that

$$\cos z = (\cos x)(\cosh y) - i(\sin x)(\sinh y) \quad \text{and} \quad |\cos z|^2 = (\cosh y)^2 - (\sin x)^2,$$

for $z = x + iy$, where x, y are real, and $\cosh y := \dfrac{e^y + e^{-y}}{2}$ and $\sinh y := \dfrac{e^y - e^{-y}}{2}$.

Exercise 1.38. We know that the equation $\cos x = 3$ has no real solution x. However, show that there are complex z that satisfy $\cos z = 3$, and find them all.

1.4.3 *Logarithm function*

In the real setting, given a positive y, $\log y \in \mathbb{R}$ is the unique real number such that $e^{\log y} = y$. Thus $\log : (0, \infty) \to \mathbb{R}$ serves as the inverse of the function $x \mapsto e^x : \mathbb{R} \to (0, \infty)$. See Figure 1.18.

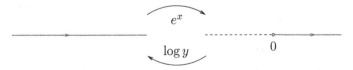

Fig. 1.18   The maps $x \mapsto e^x : \mathbb{R} \to (0, \infty)$ and $y \mapsto \log y : (0, \infty) \to \mathbb{R}$.

In the complex case, we know that $\exp : \mathbb{C} \to \mathbb{C} \setminus \{0\}$, and we now wonder if there is a "complex logarithm function" mapping $\mathbb{C} \setminus \{0\}$ to $\mathbb{C}$ that serves as an inverse to the complex exponential function. Given a $z \neq 0$, we seek a complex number w such that $\exp w = z$, and we would like to call this w the "complex logarithm of z". However, we have seen that the exponential function $\exp$ is 2π-periodic in the y-direction, and so the moment we find *one* w such that $e^w = z$, we know that there are *infinitely many* others, since $\exp(w + 2\pi i n) = \exp w = z$ for all $n \in \mathbb{Z}$. Given this infinite choice, which w must we call the complex logarithm of z? We remedy this problem of nonuniqueness by just choosing a w that lies in a *fixed* particular horizontal strip of width 2π. Indeed, all possible nonzero complex numbers can be obtained as the $\exp$ of something lying in any such strip, and now for the purpose of defining the complex logarithm, we choose (somewhat arbitrarily), the strip $\mathbb{S} := \{z \in \mathbb{C} : -\pi < \mathrm{Im}(z) \leq \pi\}$. See Figure 1.19.

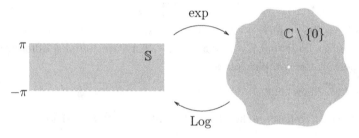

Fig. 1.19   The strip $\mathbb{S} := \mathbb{R} \times (-\pi, \pi]$ is mapped by $\exp$ onto $\mathbb{C} \setminus \{0\}$.

This will give, as we shall show below, a unique w in the strip such that $\exp w = z$, and we will call this unique w the "principal logarithm of z", denoted by $\mathrm{Log}\, z$. In order to do this, we will first introduce the notion of the principal argument of a nonzero complex number.

Principal argument of a nonzero complex number. For $z \neq 0$, let θ be the unique real number in the interval $(-\pi, \pi]$ such that

$$z = |z|(\cos\theta + i\sin\theta).$$

This value of θ is called the *principal argument of z*, and is denoted by $\mathrm{Arg}\, z$. Here are a few examples:

$$\mathrm{Arg}(3) = 0, \quad \mathrm{Arg}(-1) = \pi, \quad \mathrm{Arg}(i) = \frac{\pi}{2}, \quad \mathrm{Arg}(-i) = -\frac{\pi}{2}.$$

We note that if we start at a point on the positive real axis in the complex plane, and go around anticlockwise in a circle, then there is a sudden jump in the value of the principal argument as we cross the negative real axis: on the negative real axis, the value of the principal argument is π, while just below the negative real axis, the principal argument is close to $-\pi$.

Exercise 1.39. Depict $\left\{z \in \mathbb{C} : z \neq 0, \ \dfrac{\pi}{4} < |\mathrm{Arg}(z)| < \dfrac{\pi}{3}\right\}$ in the complex plane.

Now we are ready to define the principal logarithm of nonzero complex numbers.

Definition 1.3. The *principal logarithm* $\mathrm{Log}\,z$ of $z \neq 0$ is defined by $\mathrm{Log}\,z = \log|z| + i\mathrm{Arg}\,z$.

First of all, let us observe that

$$\exp(\mathrm{Log}\,z) = e^{\log|z|}(\cos(\mathrm{Arg}\,z) + i\sin(\mathrm{Arg}\,z))$$
$$= |z|(\cos(\mathrm{Arg}\,z) + i\sin(\mathrm{Arg}\,z)) = z.$$

This shows that $\exp : \mathbb{S} \mapsto \mathbb{C} \setminus \{0\}$ is onto. Also, it is one-to-one, because if $z_1, z_2 \in \mathbb{S}$ are such that $\exp z_1 = \exp z_2$, then $\exp(z_1 - z_2) = 1$, and so $z_1 - z_2 = 2\pi n i$, for some $n \in \mathbb{Z}$. But as $z_1, z_2 \in \mathbb{S}$, their imaginary parts must differ by a number $< 2\pi$. This implies that $n = 0$, and so $z_1 = z_2$.

So the two maps $\exp : \mathbb{S} \to \mathbb{C} \setminus \{0\}$ and $\mathrm{Log} : \mathbb{C} \setminus \{0\} \to \mathbb{S}$ are inverses of each other.

Of course, had we chosen to define the principal argument θ of a nonzero $z = |z|e^{i\theta}$ to lie in a different interval $(a, 2\pi + a]$ or $[a, 2\pi + a)$ for some other a, we would have obtained a different well-defined notion of the logarithm (which would also be equally legitimate). But when we talk about the principal logarithm of z, in this book we will *always* mean $\log|z| + i\mathrm{Arg}\,z$, with the principal argument $\mathrm{Arg}\,z \in (-\pi, \pi]$. Here is an example:

$$\mathrm{Log}(-i) = \log|-i| + i\mathrm{Arg}(-i) = \log 1 - \frac{\pi}{2}i = 0 - \frac{\pi}{2}i = -\frac{\pi}{2}i.$$

Continuity of Log in $\mathbb{C} \setminus (-\infty, 0]$. First of all, we remark that owing to the lack of continuity of $\mathrm{Arg} : \mathbb{C} \setminus \{0\} \to (-\pi, \pi]$ across the negative real axis, also the function Log is not continuous on $\mathbb{C} \setminus \{0\}$. It fails to be continuous at each point in $(-\infty, 0)$. Let us show the lack of continuity at -1. To this end, consider the sequence

$$\left(\exp\left(i\left(-\pi + \frac{1}{n}\right)\right)\right)_{n \in \mathbb{N}},$$

which converges to -1:

$$\lim_{n \to \infty} \exp\left(i\left(-\pi + \frac{1}{n}\right)\right) = \lim_{n \to \infty} (-1)\left(\cos\frac{1}{n} + i\sin\frac{1}{n}\right) = -1(1 + i0) = -1.$$

Also,

$$\text{Log}\left(\exp\left(i\left(-\pi+\frac{1}{n}\right)\right)\right) = \log 1 + i\left(-\pi+\frac{1}{n}\right) = i\left(-\pi+\frac{1}{n}\right).$$

Thus we have

$$\lim_{n\to\infty} \text{Log}\left(\exp\left(i\left(-\pi+\frac{1}{n}\right)\right)\right) = i(0-\pi) = -i\pi$$

$$\text{Log}\left(\lim_{n\to\infty} \exp\left(i\left(-\pi+\frac{1}{n}\right)\right)\right) = \text{Log}(-1) = i\pi \,,$$

showing that Log is not continuous at $-1 \in \mathbb{C} \setminus \{0\}$.

On the other hand, Log *is* continuous on the smaller set $\mathbb{C}\setminus(-\infty,0]$. This just follows by observing that the principal argument $\text{Arg}(z)$ is continuous in $\mathbb{C}\setminus(-\infty,0]$. Indeed, the key thing is that if we take any complex number z_0 not lying on $(-\infty,0]$, then there is some room around z_0 not touching the negative real axis: we can always find a small enough r such that the disc $D(z_0,r)$ does not touch the line $(-\infty,0]$. Thus, given an $\epsilon > 0$, by shrinking r further if necessary, we can ensure that the points z in $D(z_0,r)$ satisfy $|\text{Arg}(z) - \text{Arg}(z_0)| < \epsilon$. See Figure 1.20.

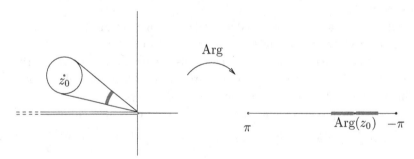

Fig. 1.20   Continuity of the principal argument Arg in $\mathbb{C} \setminus (-\infty,0]$.

As both $z \mapsto \log|z|$ and Arg are continuous in $\mathbb{C} \setminus (-\infty,0]$, it follows that Log is continuous there too.

Using the continuity of Log on $\mathbb{C}\setminus(-\infty,0]$, we will see later on that Log is complex differentiable in $\mathbb{C} \setminus (-\infty,0]$.

a^b **for** $a \in \mathbb{C} \setminus \{0\}$ **and** $b \in \mathbb{C}$. One can now also talk about a^b, where a,b are complex numbers and $a \neq 0$, and we define the *principal value of* a^b as

$$a^b := e^{b\text{Log}(a)}.$$

For example, the principal value of i^i is

$$\exp(i \cdot \text{Log}(i)) = \exp(i(\log|i| + i\text{Arg } i)) = \exp\left(i\left(0 + i\frac{\pi}{2}\right)\right) = e^{-\frac{\pi}{2}}.$$

Exercise 1.40. Find $\text{Log}(1 + i)$.

Exercise 1.41. Find $\text{Log}(-1)$ and $\text{Log}(1)$. Show that $\text{Log}(z^2)$ isn't always equal to $2 \cdot \text{Log}(z)$.

Exercise 1.42. Find the image of the annulus $\{z \in \mathbb{C} : 1 < |z| < e\}$ under the principal logarithm.

Exercise 1.43. Find the principal value of $(1 + i)^{1-i}$.

1.5 Notes

The remark on the historical development of complex numbers is taken from [Needham (1997)]. Exercise 1.2 is taken from [Shastri (2000)]. Exercises 1.7, 1.7, 1.13, 1.19 are taken from [Needham (1997)]. Exercises 1.23 and 1.34 are taken from [Beck, Marchesi, Pixton, Sabalka (2008)].

Chapter 2

Complex differentiability

In this chapter we will learn three main things:

(1) The definition of complex differentiability, that is, given $f : U \to \mathbb{C}$, where U is an open subset $\mathbb{C}$, and $z_0 \in U$, we will learn the meaning of the statement "f is complex differentiable at z_0 with complex derivative $f'(z_0)$".

(2) The Cauchy-Riemann equations: $\dfrac{\partial u}{\partial x} = \dfrac{\partial v}{\partial y}$ and $\dfrac{\partial u}{\partial y} = -\dfrac{\partial v}{\partial x}$.

These are PDEs that are satisfied by the real and imaginary parts u, v of a complex differentiable function $f : U \to \mathbb{C}$ wherever it is complex differentiable.

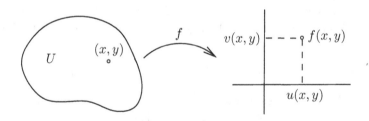

Fig. 2.1   The real and imaginary parts u, v of f.

Vice versa, we will learn that if these Cauchy-Riemann equations are satisfied everywhere in an open set and u, v are C^1, then $f = u + iv$ is complex differentiable in U.

(3) The geometric meaning of the complex derivative $f'(z_0)$: infinitesimally the map f is an amplification by $|f(z_0)|$ together with a twist (a counterclockwise rotation) through $\mathrm{Arg}(f'(z_0))$.

The central result in this chapter is the necessity and (under mild conditions) sufficiency of the Cauchy-Riemann equations for the complex differentiability of a function in an open set.

2.1 Complex differentiability

Definition 2.1.

(1) Let U be an open subset of $\mathbb{C}$, $f : U \to \mathbb{C}$ and $z_0 \in U$. Then f is said to be *complex differentiable at* z_0 if there exists a complex number L such that

$$\lim_{z \to z_0} \frac{f(z) - f(z_0)}{z - z_0} = L,$$

that is, for every $\epsilon > 0$, there is a $\delta > 0$ such that whenever $z \in U$ and $0 < |z - z_0| < \delta$, we have

$$\left| \frac{f(z) - f(z_0)}{z - z_0} - L \right| < \epsilon.$$

We denote this L (which can be shown to be unique) by

$$f'(z_0) \text{ or } \frac{df}{dz}(z_0).$$

(2) A function $f : U \to \mathbb{C}$ which is complex differentiable at all points of the open set U is called *holomorphic*[1] in U.

(3) A function that is holomorphic in $\mathbb{C}$ is called *entire*, that is, the domain of f is understood to be the whole of $\mathbb{C}$ and moreover, f is holomorphic in $\mathbb{C}$.

Let us look at a simple example of an entire function.

Example 2.1. Consider the function $f : \mathbb{C} \to \mathbb{C}$ defined by $f(z) = z^2$ ($z \in \mathbb{C}$). We show that f is entire. Note that

$$\frac{f(z) - f(z_0)}{z - z_0} = \frac{z^2 - z_0^2}{z - z_0} = z + z_0 \approx 2z_0$$

for z near z_0, and so we *guess* that $f'(z_0) = 2z_0$. Let us show this now. For $z \neq z_0$, we have

$$\left| \frac{f(z) - f(z_0)}{z - z_0} - 2z_0 \right| = \left| \frac{z^2 - z_0^2}{z - z_0} - 2z_0 \right| = |z + z_0 - 2z_0| = |z - z_0|.$$

---

[1]The word "holomorphic" is derived from the Greek "holos" meaning "entire", and "morphe" meaning "form" or "appearance".

So we see that the left hand side can be made as small as we please when z is close enough to z_0. Let $\epsilon > 0$. Set $\delta := \epsilon > 0$. Then whenever $z \in \mathbb{C}$ satisfies $0 < |z - z_0| < \delta$, we have

$$\left| \frac{f(z) - f(z_0)}{z - z_0} - 2z_0 \right| = |z - z_0| < \delta = \epsilon.$$

Hence $f'(z_0) = 2z_0$. As $z_0 \in \mathbb{C}$ was arbitrary, f is holomorphic in $\mathbb{C}$, that is, f is entire, and from the above it follows that

$$\frac{d}{dz} z^2 = 2z, \quad z \in \mathbb{C}. \qquad \lozenge$$

On the other hand, here is an example of a natural mapping which is *not* complex differentiable.

Example 2.2. Consider the function $g : \mathbb{C} \to \mathbb{C}$ defined by $g(z) = \bar{z}$ ($z \in \mathbb{C}$). We show that g is differentiable nowhere. Suppose that g is differentiable at $z_0 \in \mathbb{C}$. Let $\epsilon := \frac{1}{2} > 0$. Then there exists $\delta > 0$ such that whenever z satisfies $0 < |z - z_0| < \delta$, we have

$$\left| \frac{g(z) - g(z_0)}{z - z_0} - g'(z_0) \right| = \left| \frac{\bar{z} - \bar{z_0}}{z - z_0} - g'(z_0) \right| < \epsilon.$$

See the picture on the left of Figure 2.2. The above says that whenever z is in the punctured disc of radius δ with center z_0, we are guaranteed that this inequality holds. We will now make special choices of the z as the blue and the red point shown in the figure, and show that the special cases of the inequality above yield that $g'(z_0)$ must lie in discs of radius $1/2$ with centers at -1 and 1. See the picture on the right of Figure 2.2. But these discs have no intersection, and this will be our contradiction. We give the details below.

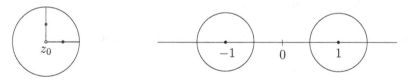

Fig. 2.2 Non complex differentiability of complex conjugation

Taking $z = z_0 + \dfrac{\delta}{2}$, we have $0 < |z - z_0| < \delta$, and so

$$\left| \frac{\bar{z} - \bar{z_0}}{z - z_0} - g'(z_0) \right| = \left| \frac{\delta/2}{\delta/2} - g'(z_0) \right| = |1 - g'(z_0)| < \epsilon. \qquad (2.1)$$

Also, taking $z = z_0 + i\dfrac{\delta}{2}$, we have $0 < |z - z_0| < \delta$, and so

$$\left|\frac{\overline{z} - \overline{z_0}}{z - z_0} - g'(z_0)\right| = \left|\frac{-i\delta/2}{i\delta/2} - g'(z_0)\right| = |1 + g'(z_0)| < \epsilon. \qquad (2.2)$$

It follows from (2.1) and (2.2) that

$$2 = |1 - g'(z_0) + 1 + g'(z_0)| \leq |1 - g'(z_0)| + |1 + g'(z_0)| < \epsilon + \epsilon = 2\epsilon = 2 \cdot \frac{1}{2} = 1,$$

a contradiction. So g is not differentiable at z_0. $\diamond$

Exercise 2.1. Show that $f : \mathbb{C} \to \mathbb{C}$ defined by $f(z) = |z|^2$ for $z \in \mathbb{C}$, is complex differentiable at 0 and that $f'(0) = 0$. We will see later (in Exercise 2.9) that f is not complex differentiable at any nonzero complex number.

Exercise 2.2. Let D be a domain, and $f : D \to \mathbb{C}$ be holomorphic in D. Set $D^* := \{z \in \mathbb{C} : \overline{z} \in D\}$, and define $f^* : D^* \to \mathbb{C}$ by $f^*(z) = \overline{f(\overline{z})}$ ($z \in D^*$). Prove that f^* is holomorphic in D^*.

The following reformulation of complex differentiability is useful to prove elementary facts about complex differentiation. Roughly speaking, the result says that for a complex differentiable function f with complex derivative L at z_0, $f(z) - f(z_0) - L \cdot (z - z_0)$ goes to 0 "faster than $z - z_0$".

Lemma 2.1. *Let U be an open set in $\mathbb{C}$, $z_0 \in U$, and $f : U \to \mathbb{C}$. Then the following are equivalent:*

(1) f *is complex differentiable at z_0 with $f'(z_0) = L$.*
(2) *There exists an $r > 0$, and a function $h : D(z_0, r) \to \mathbb{C}$, where $D(z_0, r) := \{z \in \mathbb{C} : |z - z_0| < r\}$, such that*

    (a) $f(z) = f(z_0) + (L + h(z))(z - z_0)$ *for $|z - z_0| < r$, and*
    (b) $\displaystyle\lim_{z \to z_0} h(z) = 0$.

Proof. (2)$\Rightarrow$(1): For $z \in D(z_0, r) \setminus \{z_0\}$, we have, upon rearranging, that

$$\frac{f(z) - f(z_0)}{z - z_0} - L = h(z) \overset{z \to z_0}{\longrightarrow} 0$$

and so f is complex differentiable at z_0, and $f'(z_0) = L$.

(1)$\Rightarrow$(2): Now suppose that f is complex differentiable at z_0. Then there is a $\delta_1 > 0$ such that whenever $0 < |z - z_0| < \delta_1$, $z \in U$ and

$$\left|\frac{f(z) - f(z_0)}{z - z_0} - f'(z_0)\right| < 1.$$

Set $r := \delta_1$, and define $h : D(z_0, r) \to \mathbb{C}$ by

$$h(z) = \begin{cases} \dfrac{f(z) - f(z_0)}{z - z_0} - f'(z_0) & \text{if } z \neq z_0, \\ 0 & \text{if } z = z_0. \end{cases}$$

Then $f(z) = f(z_0) + (f'(z_0) + h(z))(z - z_0)$ holds whenever $|z - z_0| < r$. If $\epsilon > 0$, then there is a $\delta > 0$ (which can be chosen smaller than r) such that whenever $0 < |z - z_0| < \delta$, we have

$$\left| \frac{f(z) - f(z_0)}{z - z_0} - f'(z_0) \right| \; (= |h(z) - 0|) < \epsilon.$$

This completes the proof. $\qquad\qquad\qquad\qquad\qquad\qquad\qquad\qquad\qquad\qquad$ $\square$

For example, using this lemma, we see that holomorphic functions must be continuous.

Exercise 2.3. Let D be a domain in $\mathbb{C}$. Show that if $f : D \to \mathbb{C}$ is complex differentiable at $z_0 \in D$, then f is continuous at z_0. Later on, we will see that if f is holomorphic in D, then in fact it is infinitely many times differentiable in D!

Using Lemma 2.1, it is also easy to show the following.

Proposition 2.1. *Let U be an open subset of $\mathbb{C}$. Let $f, g : U \to \mathbb{C}$ be complex differentiable functions at $z_0 \in U$. Then:*

(1) *$f + g$ is complex differentiable at z_0 and $(f + g)'(z_0) = f'(z_0) + g'(z_0)$.*
    (Here $f + g : U \to \mathbb{C}$ is defined by $(f + g)(z) = f(z) + g(z)$ for $z \in U$.)
(2) *If $\alpha \in \mathbb{C}$, then $\alpha \cdot f$ is complex differentiable and $(\alpha \cdot f)'(z_0) = \alpha f'(z_0)$.*
    (Here $\alpha \cdot f : U \to \mathbb{C}$ is defined by $(\alpha \cdot f)(z) = \alpha f(z)$ for $z \in U$.)
(3) *fg is complex differentiable at z_0 and moreover, there holds that $(fg)'(z_0) = f'(z_0)g(z_0) + f(z_0)g'(z_0)$.*
    (Here $fg : U \to \mathbb{C}$ is defined by $(fg)(z) = f(z)g(z)$ for $z \in U$.)

Remark 2.1. Let U be an open subset of $\mathbb{C}$, and let $\text{Hol}(U)$ denote the set of all holomorphic functions in U. Then it follows from the above that $\text{Hol}(U)$ is a complex vector space with pointwise operations. On the other hand, the third statement above shows that the pointwise product of two holomorphic functions is again holomorphic, and so $\text{Hol}(U)$ also has the structure of a ring with pointwise addition and multiplication.

Example 2.3. It is easy to see that if $f(z) := z$ $(z \in \mathbb{C})$, then $f'(z) = 1$. Using the rule for complex differentiation of a pointwise product of two holomorphic functions, it follows by induction that for all $n \in \mathbb{N}$, $z \mapsto z^n$ is entire, and

$$\frac{d}{dz}z^n = nz^{n-1}.$$

In particular all polynomials are entire. ◇

Exercise 2.4. Prove Proposition 2.1.

Exercise 2.5. Let $\mathbb{D} = \{z \in \mathbb{C} : |z| < 1\}$ and $\mathrm{Hol}(\mathbb{D})$ denote the complex vector space of all holomorphic functions in $\mathbb{D}$ with pointwise operations. Is $\mathrm{Hol}(\mathbb{D})$ finite dimensional?

Exercise 2.6. Let U be an open subset of $\mathbb{C}$, and let $f : U \to \mathbb{C}$ be such that $f(z) \neq 0$ for $z \in U$ and f is holomorphic in U. Prove that the function

$$\frac{1}{f} : U \to \mathbb{C}, \text{ defined by } \left(\frac{1}{f}\right)(z) = \frac{1}{f(z)} \text{ for all } z \in U,$$

is holomorphic, and that $\left(\dfrac{1}{f}\right)'(z) = -\dfrac{f'(z)}{(f(z))^2}$ $(z \in U)$.

Exercise 2.7. Show that in $\mathbb{C} \setminus \{0\}$, for each $m \in \mathbb{Z}$, $\dfrac{d}{dz}z^m = mz^{m-1}$.

Just like one has the chain rule for the derivative of the composition of maps in the real setting, there is an analogous chain rule in the context of composition of holomorphic functions.

Proposition 2.2. (Chain rule) *Let*

(1) D_f, D_g *be domains,*
(2) $f : D_f \to \mathbb{C}$ *be holomorphic in* D_f,
(3) $g : D_g \to \mathbb{C}$ *be holomorphic in* D_g, *and*
(4) $f(D_f) \subset D_g$.

Then their composition $g \circ f : D_f \to \mathbb{C}$, *defined by* $(g \circ f)(z) = g(f(z))$, $z \in D_f$, *is holomorphic in* D_f *and* $(g \circ f)'(z) = g'(f(z))f'(z)$ *for all* $z \in D_f$.

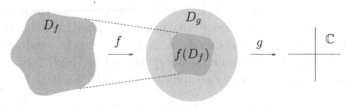

Fig. 2.3   The chain rule: $(g \circ f)'(z) = g'(f(z))f'(z)$, $z \in D_f$.

Proof. Let $z_0 \in D_f$. Then $f(z_0) \in D_g$. From the complex differentiability of f at z_0 and that of g at $f(z_0)$, we know that there are functions h_f and h_g, defined in the discs $D(z_0, r_f) \subset D_f$ and $D(f(z_0), r_g) \subset D_g$ such that

$$f(z) - f(z_0) = (f'(z_0) + h_f(z))(z - z_0),$$
$$g(w) - g(f(z_0)) = (g'(f(z_0)) + h_g(w))(w - f(z_0)),$$

and

$$\lim_{z \to z_0} h_f(z) = 0, \quad \lim_{w \to f(z_0)} h_g(w) = 0.$$

But it follows from the continuity of f at z_0 that when z is close to z_0, $w := f(z)$ is close to $f(z_0)$, and so if $z \neq z_0$, but close to z_0, we have

$$\frac{(g \circ f)(z) - (g \circ f)(z_0)}{z - z_0} = (g'(f(z_0)) + h_g(f(z)))(f'(z_0) + h_f(z)),$$

and so the claim follows. $\qquad\qquad\square$

Example 2.4. From Exercise 2.7, we know that

$$\frac{d}{dz}\left(\frac{1}{z}\right) = -\frac{1}{z^2}, \quad z \in \mathbb{C} \setminus \{0\},$$

but this is also easy to see from the definition, because

$$\frac{\frac{1}{z} - \frac{1}{z_0}}{z - z_0} = \frac{z_0 - z}{z z_0 (z - z_0)} = \frac{-1}{z z_0} \xrightarrow{z \to z_0} -\frac{1}{z_0^2}$$

for $z_0 \in \mathbb{C} \setminus \{0\}$.

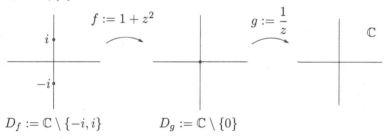

$D_f := \mathbb{C} \setminus \{-i, i\}$ $\qquad\qquad$ $D_g := \mathbb{C} \setminus \{0\}$

Fig. 2.4 Application of the chain rule.

Now consider the functions $f := 1 + z^2$ on $D_f := \mathbb{C} \setminus \{-i, i\}$ and $g := 1/z$ on $D_g := \mathbb{C} \setminus \{0\}$. Clearly, $f(D_f) \subset D_g$, and so, by the Chain rule,

$$\frac{d}{dz}\left(\frac{1}{1 + z^2}\right) = -\frac{1}{(1 + z^2)^2} \cdot 2z = -\frac{2z}{(1 + z^2)^2}$$

in $\mathbb{C} \setminus \{-i, i\}$. $\qquad\qquad\Diamond$

Exercise 2.8. Assuming that $\exp z$ is entire and that $\exp' z = \exp z$ (we will prove this later), show that

$$z \mapsto \exp\left(-\frac{1+z}{1-z}\right)$$

is holomorphic in the unit disc $\mathbb{D} := \{z \in \mathbb{C} : |z| < 1\}$, and find its derivative.

2.2 Cauchy-Riemann equations

We now prove the main result in this chapter, which says roughly that a function $f = u + iv$ is holomorphic if and only if its real and imaginary parts u, v (viewed as real valued functions living in an open subset of $\mathbb{R}^2$) satisfy a pair of partial differential equations, called the Cauchy-Riemann equations.

Cauchy-Riemann

Let U be an open subset of $\mathbb{C}$, and let $f : U \to \mathbb{C}$ be a function. Then taking any point $(x, y) \in U$, we have $f(x + iy) \in \mathbb{C}$, and we can look at the real part $u(x, y)$ of $f(x + iy)$, and the imaginary part $v(x, y)$ of $f(x + iy)$. See Figure 2.5.

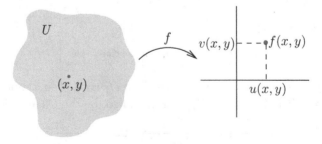

Fig. 2.5 The real and imaginary parts u, v of f.

If one changes the point (x, y), then $f(x + iy)$ changes, and so do $u(x, y)$ and $v(x, y)$. In this manner, associated with f, we obtain two *real-valued* functions

$$u : U \to \mathbb{R}, \quad U \ni (x, y) \mapsto \mathrm{Re}(f(x + iy)) =: u(x, y),$$
$$v : U \to \mathbb{R}, \quad U \ni (x, y) \mapsto \mathrm{Im}(f(x + iy)) =: v(x, y).$$

Our first result in this section is the necessity of the Cauchy-Riemann equations for complex differentiability, and we will prove this result in Theorem 2.1 below. The result says that if f is complex differentiable at $(x_0, y_0) \in U$, then

$$\boxed{\frac{\partial u}{\partial x} = \frac{\partial v}{\partial y} \quad \text{and} \quad \frac{\partial u}{\partial y} = -\frac{\partial v}{\partial x}} \quad \text{at } (x_0, y_0),$$

and these two equations are called the Cauchy-Riemann equations. So for a function to be complex differentiable, it has got to satisfy these equations. In other words, if these equations aren't satisfied by the real and imaginary part of some complex valued function at some point, then at that point we know that the function can't be complex differentiable. Here is an example. We had seen earlier (by "brute force", using the ϵ-δ definition of complex differentiability), that $z \mapsto \bar{z}$ is not complex differentiable anywhere in the complex plane. Let us revisit Example 2.2.

Example 2.5. For the function $g : \mathbb{C} \to \mathbb{C}$ defined by $g(z) = \bar{z}$ ($z \in \mathbb{C}$), we have that

$$u(x, y) = \mathrm{Re}(g(x + iy)) = \mathrm{Re}(x - iy) = x,$$
$$v(x, y) = \mathrm{Im}(g(x + iy)) = \mathrm{Im}(x - iy) = -y.$$

Thus

$$\frac{\partial u}{\partial x}(x, y) = 1 \neq -1 = \frac{\partial v}{\partial y}(x, y).$$

This shows that the Cauchy-Riemann equations can't hold at any point. So we recover our previous observation that g is complex differentiable nowhere. $\Diamond$

Before proving the necessity of the Cauchy-Riemann equations for complex differentiability, let us also mention the second important result we will show in this section, namely the sufficiency of the Cauchy-Riemann equations for holomorphicity in an open set. More precisely, we will show in

Theorem 2.2 below that if $u, v : U \to \mathbb{R}$ are two real-valued functions in an open set U that are continuously differentiable in U (as functions of two real variables), and the Cauchy-Riemann equations are satisfied by u and v everywhere in U, then the new *complex-valued* function $f : U \to \mathbb{C}$, created from u, v by setting $f(x + iy) := u(x, y) + iv(x, y)$, $(x, y) \in U$, (so that u, v turn out to be the real and imaginary parts, respectively, of the f we have constructed), is holomorphic in U. This important result will enable us to establish the holomorphicity of important functions without having to go through the rigmarole of verifying the ϵ-δ definition. Here is an example.

Example 2.6. Consider the functions u, v defined in the punctured plane $\mathbb{R}^2 \setminus \{(0,0)\}$ as follows:

$$u(x, y) := \frac{x}{x^2 + y^2}, \quad v(x, y) := \frac{-y}{x^2 + y^2}, \quad (x, y) \neq (0, 0).$$

Then we have

$$\frac{\partial u}{\partial x} = \frac{1 \cdot (x^2 + y^2) - x \cdot 2x}{(x^2 + y^2)^2} = \frac{y^2 - x^2}{(x^2 + y^2)^2},$$

$$\frac{\partial u}{\partial y} = \frac{-x \cdot 2y}{(x^2 + y^2)^2} = \frac{-2xy}{(x^2 + y^2)^2},$$

$$\frac{\partial v}{\partial x} = \frac{2xy}{(x^2 + y^2)^2}, \quad \text{and} \quad \frac{\partial v}{\partial y} = \frac{y^2 - x^2}{(x^2 + y^2)^2}.$$

Clearly $(x, y) \mapsto (x^2 + y^2)^2, y^2 - x^2, \pm 2xy$ are continuous in $\mathbb{R}^2$ and $(x^2 + y^2)^2$ is nonzero in $\mathbb{R}^2 \setminus \{(0,0)\}$. So it follows that each of these partial derivatives is continuous in $\mathbb{R}^2 \setminus \{(0,0)\}$. So u, v are continuously differentiable in $\mathbb{R}^2 \setminus \{(0,0)\}$. Also the Cauchy-Riemann equations hold. Thus $f := u + iv$ is holomorphic in $\mathbb{C} \setminus \{0\}$.

In fact, the f above is just the reciprocation map $z \mapsto 1/z$:

$$f = u + iv = \frac{x}{x^2 + y^2} + i\left(\frac{-y}{x^2 + y^2}\right) = \frac{x - iy}{x^2 + y^2} = \frac{\bar{z}}{|z|^2} = \frac{\bar{z}}{z\bar{z}} = \frac{1}{z}, \quad z \neq 0.$$

$\diamond$

Theorem 2.1. *Let U be an open subset of $\mathbb{C}$ and let $f : U \to \mathbb{C}$ be complex differentiable at $z_0 = x_0 + iy_0 \in U$. Then the functions*

$$(x, y) \mapsto u(x, y) := \mathrm{Re}(f(x + iy)) : U \to \mathbb{R} \ and$$

$$(x, y) \mapsto v(x, y) := \mathrm{Im}(f(x + iy)) : U \to \mathbb{R}$$

are differentiable at (x_0, y_0) and

$$\frac{\partial u}{\partial x}(x_0, y_0) = \frac{\partial v}{\partial y}(x_0, y_0) \quad and \quad \frac{\partial u}{\partial y}(x_0, y_0) = -\frac{\partial v}{\partial x}(x_0, y_0). \tag{2.3}$$

Proof.   (The idea of the proof is easy, we just let (x, y) tend to (x_0, y_0) by first keeping y fixed at y_0, and then by keeping x fixed at x_0, and look at what this gives us. See Figure 2.6.)

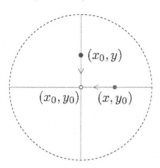

Fig. 2.6   Proof of the necessity of the Cauchy-Riemann (CR) equations for complex differentiability.

Let $z_0 = (x_0, y_0) \in U$. Let $\epsilon > 0$. Then there is $\delta > 0$ such that whenever $0 < |z - z_0| < \delta$, we have $z \in U$ and

$$\left| \frac{f(z) - f(z_0)}{z - z_0} - f'(z_0) \right| < \epsilon. \tag{2.4}$$

Step 1. We will show that $\dfrac{\partial u}{\partial x}(x_0, y_0)$ exists and equals $\operatorname{Re}(f'(z_0))$.

Let $z := x + iy_0$, where $x \in \mathbb{R}$ is such that $0 < |x - x_0| < \delta$. Then $z - z_0 = x - x_0$, and so $0 < |z - z_0| = |x - x_0| < \delta$. Thus

$$
\begin{aligned}
\left| \frac{u(x, y_0) - u(x_0, y_0)}{x - x_0} \right. & \left. - \operatorname{Re}(f'(z_0)) \right| \\
&= \left| \operatorname{Re}\left(\frac{f(x + iy_0) - f(x_0 + iy_0)}{x - x_0} \right) - \operatorname{Re}(f'(z_0)) \right| \\
&= \left| \operatorname{Re}\left(\frac{f(z) - f(z_0)}{z - z_0} \right) - \operatorname{Re}(f'(z_0)) \right| \\
&\leq \left| \frac{f(z) - f(z_0)}{z - z_0} - f'(z_0) \right| < \epsilon,
\end{aligned}
$$

using (2.4). Thus the partial derivative

$$
\begin{aligned}
\frac{\partial u}{\partial x}(x_0, y_0) &= \lim_{x \to x_0} \frac{u(x, y_0) - u(x_0, y_0)}{x - x_0} \\
&= \operatorname{Re}(f'(z_0)).
\end{aligned}
$$

Step 2. We show that $\dfrac{\partial v}{\partial x}(x_0, y_0) = \mathrm{Im}(f'(z_0))$.

Proceeding in a manner similar to Step 1 above, we also have with the same notation that

$$\left| \frac{v(x, y_0) - v(x_0, y_0)}{x - x_0} - \mathrm{Im}(f'(z_0)) \right|$$

$$= \left| \mathrm{Im}\left(\frac{f(x + iy_0) - f(x_0 + iy_0)}{x - x_0} \right) - \mathrm{Im}(f'(z_0)) \right|$$

$$= \left| \mathrm{Im}\left(\frac{f(z) - f(z_0)}{z - z_0} \right) - \mathrm{Im}(f'(z_0)) \right|$$

$$\leq \left| \frac{f(z) - f(z_0)}{z - z_0} - f'(z_0) \right| < \epsilon,$$

and so $\dfrac{\partial v}{\partial x}(x_0, y_0) = \lim\limits_{x \to x_0} \dfrac{v(x, y_0) - v(x_0, y_0)}{x - x_0} = \mathrm{Im}(f'(z_0))$. Thus

$$f'(z_0) = \frac{\partial u}{\partial x}(x_0, y_0) + i\frac{\partial v}{\partial x}(x_0, y_0). \tag{2.5}$$

Step 3. We show that $\dfrac{\partial u}{\partial y}(x_0, y_0) = -\mathrm{Im}(f'(z_0))$.

Now let $z := x_0 + iy$, where $y \in \mathbb{R}$ is such that $0 < |y - y_0| < \delta$. Then $z - z_0 = i(y - y_0)$ and so $0 < |z - z_0| = |y - y_0| < \delta$. Thus using the fact that for $a, b \in \mathbb{R}$ there holds that $\mathrm{Re}(a + ib) = \mathrm{Im}(i(a + ib))$, we obtain that

$$\left| \frac{u(x_0, y) - u(x_0, y_0)}{y - y_0} + \mathrm{Im}(f'(z_0)) \right| = \left| \frac{\mathrm{Im}(i(f(z) - f(z_0)))}{y - y_0} + \mathrm{Im}(f'(z_0)) \right|$$

$$= \left| \mathrm{Im}\left(-\frac{f(z) - f(z_0)}{z - z_0} + f'(z_0) \right) \right|$$

$$\leq \left| \frac{f(z) - f(z_0)}{z - z_0} - f'(z_0) \right| < \epsilon.$$

Thus the partial derivative

$$\frac{\partial u}{\partial y}(x_0, y_0) = \lim_{y \to y_0} \frac{u(x_0, y) - u(x_0, y_0)}{y - y_0} = -\mathrm{Im}(f'(z_0)).$$

Recall that in Step 2, we had obtained

$$\frac{\partial v}{\partial x}(x_0, y_0) = \mathrm{Im}(f'(z_0)),$$

and so, we have established one of the two Cauchy-Riemann equations, namely that

$$\frac{\partial u}{\partial y}(x_0, y_0) = -\frac{\partial v}{\partial x}(x_0, y_0).$$

Step 4. We show that $\dfrac{\partial v}{\partial y}(x_0, y_0) = \text{Re}(f'(z_0))$.

Proceeding as in Step 3 above, with the same notation, and using the fact that for $a, b \in \mathbb{R}$ there holds that $\text{Im}(a + ib) = -\text{Re}(i(a + ib))$, we have that

$$\left| \frac{v(x_0, y) - v(x_0, y_0)}{y - y_0} - \text{Re}(f'(z_0)) \right| = \left| -\text{Re}\left(i \frac{f(z) - f(z_0)}{y - y_0} \right) - \text{Re}(f'(z_0)) \right|$$

$$\leq \left| -i \frac{f(z) - f(z_0)}{y - y_0} - f'(z_0) \right|$$

$$= \left| \frac{f(z) - f(z_0)}{z - z_0} - f'(z_0) \right| < \epsilon.$$

Thus the partial derivative

$$\frac{\partial v}{\partial y}(x_0, y_0) = \lim_{y \to y_0} \frac{v(x_0, y) - v(x_0, y_0)}{y - y_0} = \text{Re}(f'(z_0)).$$

Hence

$$f'(z_0) = \frac{\partial v}{\partial y}(x_0, y_0) - i \frac{\partial u}{\partial y}(x_0, y_0). \tag{2.6}$$

From (2.5) and (2.6), it follows that

$$\frac{\partial u}{\partial x}(x_0, y_0) = \frac{\partial v}{\partial y}(x_0, y_0) \quad \text{and} \quad \frac{\partial v}{\partial x}(x_0, y_0) = -\frac{\partial u}{\partial y}(x_0, y_0).$$

So we have got both Cauchy-Riemann equations.

Finally, we show that u, v are differentiable (as real valued functions of two real variables) at (x_0, y_0). For $z = (x, y)$ satisfying $0 < |z - z_0| < \delta$,

$$\frac{\left| u(x, y) - u(x_0, y_0) - \left(\dfrac{\partial u}{\partial x}(x_0, y_0) \right)(x - x_0) - \left(\dfrac{\partial u}{\partial y}(x_0, y_0) \right)(y - y_0) \right|}{\|(x, y) - (x_0, y_0)\|_2}$$

$$= \frac{\left| u(x, y) - u(x_0, y_0) - \left(\dfrac{\partial u}{\partial x}(x_0, y_0) \right)(x - x_0) + \left(\dfrac{\partial v}{\partial x}(x_0, y_0) \right)(y - y_0) \right|}{\|(x, y) - (x_0, y_0)\|_2}$$

$$= \frac{|\text{Re}(f(z) - f(z_0) - f'(z_0)(z - z_0))|}{|z - z_0|} < \epsilon.$$

Thus u is differentiable at (x_0, y_0). Similarly, v is also differentiable at (x_0, y_0). $\qquad\square$

Remark 2.2. We will see later on that in fact the real and imaginary parts of a holomorphic function are infinitely many times differentiable.

Exercise 2.9. Consider Exercise 2.1 again. Show that f is not differentiable at any point of the open set $\mathbb{C} \setminus \{0\}$.

One also has the following converse to Theorem 2.1. This is a very useful result to check the holomorphicity of functions.

Theorem 2.2. *Let*

(1) *U be an open subset of $\mathbb{C}$,*
(2) *$u, v : U \to \mathbb{R}$ be continuously differentiable, and*
(3) *u, v satisfy the Cauchy-Riemann equations:*

$$\text{for all } (x,y) \in U, \quad \frac{\partial u}{\partial x}(x,y) = \frac{\partial v}{\partial y}(x,y) \text{ and } \frac{\partial u}{\partial y}(x,y) = -\frac{\partial v}{\partial x}(x,y).$$

Then $f := u + iv : U \to \mathbb{C}$ is holomorphic in U and

$$f'(x+iy) = \frac{\partial u}{\partial x}(x,y) + i\frac{\partial v}{\partial x}(x,y) \text{ for } x+iy \in U.$$

Proof. Let $z_0 = x_0 + iy_0 \in U$. Let $\epsilon > 0$. Let $\delta > 0$ be such that whenever $z = x + iy$ belongs to the disc $D(z_0, \delta) := \{w \in \mathbb{C} : |w - z_0| < \delta\}$, we have $z \in U$,

$$\left|\frac{\partial u}{\partial x}(x,y) - \frac{\partial u}{\partial x}(x_0, y_0)\right| < \epsilon, \text{ and } \left|\frac{\partial v}{\partial x}(x,y) - \frac{\partial v}{\partial x}(x_0, y_0)\right| < \epsilon. \quad (2.7)$$

(That this is possible follows from the fact that u, v are continuously differentiable.)

Let $z = x + iy$ be a fixed point in the punctured disc $D(z_0, \delta) \setminus \{z_0\}$, and consider the part of the line in $D(z_0, \delta)$ joining z_0 to z. A point on this line can be represented by

$$p(t) = (1-t)z_0 + tz = \Big((1-t)x_0 + tx, \ (1-t)y_0 + ty\Big).$$

In particular, $p(0) = z_0$ and $p(1) = z$. See Figure 2.7.

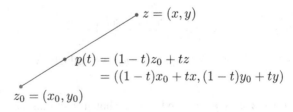

Fig. 2.7   Points $p(t)$, on the line joining z_0 to z.

Define $\varphi_1, \varphi_2 : (-1, 1) \to \mathbb{R}$ by

$$\begin{bmatrix} \varphi_1(t) \\ \varphi_2(t) \end{bmatrix} := \begin{bmatrix} u(p(t)) \\ v(p(t)) \end{bmatrix}.$$

Then using the chain rule we obtain

$$\begin{bmatrix} \varphi_1'(t) \\ \varphi_2'(t) \end{bmatrix} := \begin{bmatrix} \dfrac{\partial u}{\partial x}(p(t)) \cdot (x - x_0) + \dfrac{\partial u}{\partial y}(p(t)) \cdot (y - y_0) \\ \dfrac{\partial v}{\partial x}(p(t)) \cdot (x - x_0) + \dfrac{\partial v}{\partial y}(p(t)) \cdot (y - y_0) \end{bmatrix}.$$

Let

$$A(t) := \frac{\partial u}{\partial x}(p(t)) = \frac{\partial v}{\partial y}(p(t)),$$

$$B(t) := -\frac{\partial u}{\partial y}(p(t)) = \frac{\partial v}{\partial x}(p(t)),$$

where we have used the Cauchy-Riemann equations to obtain the rightmost equalities. Thus with this notation, we have

$$\begin{bmatrix} \varphi_1'(t) \\ \varphi_2'(t) \end{bmatrix} = \begin{bmatrix} A(t)(x - x_0) - B(t)(y - y_0) \\ B(t)(x - x_0) + A(t)(y - y_0) \end{bmatrix} = \begin{bmatrix} \mathrm{Re}((A(t) + iB(t))(z - z_0)) \\ \mathrm{Im}((A(t) + iB(t))(z - z_0)) \end{bmatrix}.$$

So

$$f(z) - f(z_0) = u(x, y) - u(x_0, y_0) + i(v(x, y) - v(x_0, y_0))$$

$$= \varphi_1(1) - \varphi_1(0) + i(\varphi_2(1) - \varphi_2(0))$$

$$= \int_0^1 \varphi_1'(t)dt + i \int_0^1 \varphi_2'(t)dt$$

$$= \left(\int_0^1 A(t)dt + i \int_0^1 B(t)dt \right) \cdot (z - z_0).$$

Thus

$$\frac{f(z) - f(z_0)}{z - z_0} - \left(\frac{\partial u}{\partial x}(x_0, y_0) + i\frac{\partial v}{\partial x}(x_0, y_0) \right)$$

$$= \int_0^1 \left(\frac{\partial u}{\partial x}(p(t)) - \frac{\partial u}{\partial x}(p(0)) \right) dt + i \int_0^1 \left(\frac{\partial v}{\partial x}(p(t)) - \frac{\partial v}{\partial x}(p(0)) \right) dt.$$

By (2.7), it follows that

$$\left| \frac{f(z) - f(z_0)}{z - z_0} - \left(\frac{\partial u}{\partial x}(x_0, y_0) + i\frac{\partial v}{\partial x}(x_0, y_0) \right) \right| < \epsilon + \epsilon = 2\epsilon.$$

This holds for all z satisfying $0 < |z - z_0| < \delta$, and so f is complex differentiable at z_0 and

$$f'(z_0) = \frac{\partial u}{\partial x}(x_0, y_0) + i\frac{\partial v}{\partial x}(x_0, y_0).$$

This completes the proof. $\qquad\square$

Let us revisit Example 2.1 again, but now instead of using the ϵ-δ definition of complex differentiability, we will use the above result to check holomorphicity of the squaring map.

Example 2.7. For the function $f : \mathbb{C} \to \mathbb{C}$ defined by $f(z) = z^2$ $(z \in \mathbb{C})$ we have that

$$u(x,y) = \mathrm{Re}(f(x+iy)) = \mathrm{Re}(x^2 - y^2 + 2xyi) = x^2 - y^2,$$
$$v(x,y) = \mathrm{Im}(f(x+iy)) = \mathrm{Im}(x^2 - y^2 + 2xyi) = 2xy.$$

Thus

$$\frac{\partial u}{\partial x}(x,y) = 2x = \frac{\partial v}{\partial y}(x,y),$$
$$\frac{\partial u}{\partial y}(x,y) = -2y = -\frac{\partial v}{\partial x}(x,y),$$

which shows that the Cauchy-Riemann equations hold in $\mathbb{C}$. So we recover our previous observation that f is entire, and since

$$f'(z) = \frac{\partial u}{\partial x}(x,y) + i\frac{\partial v}{\partial x}(x,y) = 2x + 2yi = 2z,$$

we also obtain that $f'(z) = 2z$ for $z \in \mathbb{C}$. $\Diamond$

Example 2.8. (exp, sin, cos are entire.) For the function $g : \mathbb{C} \to \mathbb{C}$ defined by $g(z) = \exp z$ $(z \in \mathbb{C})$ we have that

$$u(x,y) = \mathrm{Re}(g(x+iy)) = \mathrm{Re}(e^x(\cos y + i\sin y)) = e^x \cos y,$$
$$v(x,y) = \mathrm{Im}(g(x+iy)) = \mathrm{Im}(e^x(\cos y + i\sin y)) = e^x \sin y.$$

Thus

$$\frac{\partial u}{\partial x}(x,y) = e^x \cos y = \frac{\partial v}{\partial y}(x,y),$$
$$\frac{\partial u}{\partial y}(x,y) = -e^x \sin y = -\frac{\partial v}{\partial x}(x,y),$$

which shows that the Cauchy-Riemann equations hold in $\mathbb{C}$. So we arrive at the important result that exp is entire, and since

$$g'(z) = \frac{\partial u}{\partial x}(x,y) + i\frac{\partial v}{\partial x}(x,y) = e^x \cos y + ie^x \sin y = \exp z,$$

we also obtain that $\dfrac{d}{dz}\exp z = \exp z$ for $z \in \mathbb{C}$. Hence from Proposition 2.2, also the trigonometric functions

$$\sin z = \frac{\exp(iz) - \exp(-iz)}{2i} \quad \text{and} \quad \cos z = \frac{\exp(iz) + \exp(-iz)}{2}$$

are entire functions, and moreover,

$$\frac{d}{dz}\sin z = \frac{i\exp(iz) - (-i)\exp(-iz)}{2i} = \frac{\exp(iz) + \exp(-iz)}{2} = \cos z, \text{ and}$$

$$\frac{d}{dz}\cos z = \frac{i\exp(iz) + (-i)\exp(-iz)}{2} = -\frac{\exp(iz) - \exp(-iz)}{2i} = -\sin z.$$

$$\diamondsuit$$

Example 2.9. (Holomorphicity of Log.) We will show that the principal logarithm is holomorphic in the open set $\mathbb{C}\backslash(-\infty, 0]$. Note that the principal logarithm is defined in the bigger set $\mathbb{C}\backslash\{0\}$, but we had seen earlier that it is not continuous in this bigger set (because at each negative real number, it is discontinuous). We had also seen that in the smaller set $\mathbb{C}\backslash(-\infty, 0]$, the principal logarithm is continuous. We will now use this continuity to show that Log is in fact holomorphic in $\mathbb{C}\backslash(-\infty, 0]$, and that

$$\frac{d}{dz}\mathrm{Log}(z) = \frac{1}{z} \quad \text{for } z \in \mathbb{C}\backslash(-\infty, 0].$$

First note that when $z, z_0 \in \mathbb{C}\backslash(-\infty, 0]$ are distinct, then $\mathrm{Log}(z) \neq \mathrm{Log}(z_0)$. (Why?) Let $\epsilon > 0$. Set

$$\epsilon_1 := \min\left\{\frac{|z_0|}{2}, \frac{|z_0|^2}{2}\epsilon\right\}.$$

Since $\exp w$ is differentiable at $w_0 := \mathrm{Log}(z_0)$, there is a $\delta_1 > 0$ such that whenever w satisfies $0 < |w - w_0| = |w - \mathrm{Log}(z_0)| < \delta_1$, there holds that

$$\left|\frac{\exp w - \exp w_0}{w - w_0} - \exp w_0\right| = \left|\frac{\exp w - z_0}{w - \mathrm{Log}(z_0)} - z_0\right| < \epsilon_1.$$

But by the continuity and injectivity of Log in $\mathbb{C}\backslash(-\infty, 0]$, there exists a $\delta > 0$ such that whenever $0 < |z - z_0| < \delta$, we have

$$0 < |\mathrm{Log}\, z - \mathrm{Log}\, z_0| < \delta_1.$$

Thus with $w := \mathrm{Log}\, z$, and $0 < |z - z_0| < \delta$, we have $0 < |w - w_0| < \delta_1$, and so

$$\left|\frac{z - z_0}{\mathrm{Log}\, z - \mathrm{Log}\, z_0} - z_0\right| < \epsilon_1.$$

But then $\left| \dfrac{z - z_0}{\operatorname{Log} z - \operatorname{Log} z_0} \right| \geq |z_0| - \epsilon_1 \geq \dfrac{|z_0|}{2}$. So whenever $0 < |z - z_0| < \delta$, we have

$$\left| \frac{\operatorname{Log} z - \operatorname{Log} z_0}{z - z_0} - \frac{1}{z_0} \right| = \left| \left(z_0 - \frac{z - z_0}{\operatorname{Log} z - \operatorname{Log} z_0} \right) \cdot \frac{1}{\dfrac{z - z_0}{\operatorname{Log} z - \operatorname{Log} z_0}} \cdot \frac{1}{z_0} \right|$$

$$= \left| z_0 - \frac{z - z_0}{\operatorname{Log} z - \operatorname{Log} z_0} \right| \cdot \frac{1}{\left| \dfrac{z - z_0}{\operatorname{Log} z - \operatorname{Log} z_0} \right|} \cdot \frac{1}{|z_0|}$$

$$< \epsilon_1 \cdot \frac{1}{|z_0|/2} \cdot \frac{1}{|z_0|} = \frac{2\epsilon_1}{|z_0|^2} < \epsilon.$$

Thus Log is holomorphic in $\mathbb{C} \setminus (-\infty, 0]$ and moreover, $\dfrac{d}{dz} \operatorname{Log} z = \dfrac{1}{z}$. $\lozenge$

We now consider an example illustrating the fact that the assumption of differentiability of u, v in Theorem 2.2 (as opposed to mere existence of partial derivatives of u, v satisfying the Cauchy-Riemann equations), is not superfluous.

Example 2.10. Consider the function $f : \mathbb{C} \to \mathbb{C}$ given by

$$f(x + iy) = \frac{xy(x + iy)}{x^2 + y^2}$$

if $x + iy \neq 0$, and $f(0) = 0$. We have that for nonzero $(x, y) \in \mathbb{R}^2$,

$$u(x, y) = \operatorname{Re}(f(x + iy)) = \frac{x^2 y}{x^2 + y^2},$$

$$v(x, y) = \operatorname{Im}(f(x + iy)) = \frac{xy^2}{x^2 + y^2},$$

and $u(0, 0) = v(0, 0) = 0$. Thus

$$\frac{\partial u}{\partial x}(0, 0) = 0 = \frac{\partial v}{\partial y}(0, 0), \text{ and } \frac{\partial u}{\partial y}(0, 0) = 0 = -\frac{\partial v}{\partial x}(0, 0),$$

which shows that the Cauchy-Riemann equations hold at the point $(0, 0)$. However, the function is not complex differentiable at 0, since if it were, we would have

$$f'(0) = \frac{\partial u}{\partial x}(0, 0) + i \frac{\partial v}{\partial x}(0, 0) = 0 + i0 = 0,$$

and for $\epsilon = 1/4$, there would exist a corresponding δ such that whenever $0 < |z - 0| = |x + iy| < \delta$, we would have

$$\left| \frac{f(z) - f(0)}{z - 0} - f'(0) \right| = \left| \frac{xy}{x^2 + y^2} \right| < \epsilon,$$

but taking $x + iy = \dfrac{\delta}{2} + i\dfrac{\delta}{2}$, we arrive at the contradiction that

$$\frac{1}{2} = \left| \frac{xy}{x^2 + y^2} \right| < \epsilon = \frac{1}{4}.$$

This shows that f is not complex differentiable at 0.

We note that there is no contradiction to Theorem 2.2, since for example u is not differentiable at $(0,0)$. It it were, its derivative at $(0,0)$ would have to be the linear transformation

$$\begin{bmatrix} x \\ y \end{bmatrix} \mapsto \begin{bmatrix} \dfrac{\partial u}{\partial x}(0,0) & \dfrac{\partial u}{\partial y}(0,0) \end{bmatrix} \begin{bmatrix} x \\ y \end{bmatrix} = \begin{bmatrix} 0 & 0 \end{bmatrix} \begin{bmatrix} x \\ y \end{bmatrix} = 0.$$

But then with $\epsilon := 1/3 > 0$, there must exist a $\delta > 0$ such that whenever $0 < \|(x,y) - (0,0)\|_2 < \delta$, we would have

$$\frac{|u(x,y) - u(0,0) - 0((x,y) - (0,0))|}{\|(x,y) - (0,0)\|_2} = \frac{x^2 y}{(x^2 + y^2)^{\frac{3}{2}}} < \epsilon = \frac{1}{3}.$$

Then with $(x,y) = \left(\dfrac{\delta}{2}, \dfrac{\delta}{2} \right)$, we have $\|(x,y) - (0,0)\|_2 = \dfrac{\delta}{\sqrt{2}} < \delta$, and so

$$\frac{x^2 y}{(x^2 + y^2)^{\frac{3}{2}}} = \frac{\dfrac{\delta^2}{4} \cdot \dfrac{\delta}{2}}{\left(\dfrac{\delta^2}{4} + \dfrac{\delta^2}{4} \right)^{\frac{3}{2}}} = \frac{1}{\sqrt{8}} < \epsilon = \frac{1}{3} = \frac{1}{\sqrt{9}},$$

a contradiction. So u is not differentiable at $(0,0)$. $\Diamond$

The Cauchy-Riemann equations can also be used to prove some interesting facts, for example the following one, which highlights the "rigidity" of holomorphic functions alluded to earlier. See also Exercise 2.12 below.

Example 2.11. (Holomorphic function with constant modulus on a disc is a constant.) Consider the disc $D = \{z \in \mathbb{C} : |z - z_0| < r\}$. We will show using the Cauchy-Riemann equations that if $f : D \to \mathbb{C}$ is holomorphic in D, with the property that there is a $c \in \mathbb{R}$ such that $|f(z)| = c$ for all $z \in D$, then f is constant. (We will use this fact in later when we learn about a result called the "Maximum Modulus Theorem".) See the picture below.

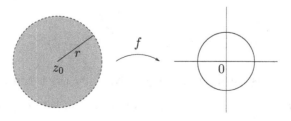

Let u, v denote the real and imaginary parts of f. By assumption, we have $c^2 = |f|^2 = u^2 + v^2$, and so by differentiation,

$$u\frac{\partial u}{\partial x} + v\frac{\partial v}{\partial x} = 0,$$

$$u\frac{\partial u}{\partial y} + v\frac{\partial v}{\partial y} = 0.$$

Using $\dfrac{\partial v}{\partial x} = -\dfrac{\partial u}{\partial y}$ in the first equation and $\dfrac{\partial v}{\partial y} = \dfrac{\partial u}{\partial x}$ in the second equation,

$$u\frac{\partial u}{\partial x} - v\frac{\partial u}{\partial y} = 0, \tag{2.8}$$

$$u\frac{\partial u}{\partial y} + v\frac{\partial u}{\partial x} = 0. \tag{2.9}$$

To remove $\dfrac{\partial u}{\partial y}$, multiply (2.8) by u, (2.9) by v, and add: $(u^2 + v^2)\dfrac{\partial u}{\partial x} = 0.$

To remove $\dfrac{\partial u}{\partial x}$, multiply (2.8) by $-v$, (2.9) by u, and add: $(u^2 + v^2)\dfrac{\partial u}{\partial y} = 0.$

If $c = 0$ so that $u^2 + v^2 = c^2 = 0$, then $u = v = 0$ and so $f = 0$ in D.

If $c \neq 0$, then $\dfrac{\partial u}{\partial x} = \dfrac{\partial u}{\partial y} = 0$, and the CR equations give also $\dfrac{\partial v}{\partial x} = \dfrac{\partial v}{\partial y} = 0.$

By the Fundamental Theorem of Integral Calculus, it follows that

$$u(x, y_0) - u(x_0, y_0) = \int_{x_0}^{x} \frac{\partial u}{\partial x}(\xi, y_0)d\xi = 0,$$

$$u(x, y) - u(x, y_0) = \int_{y_0}^{y} \frac{\partial u}{\partial y}(x, \eta)d\eta = 0.$$

So the value of u at any (x, y) is the same as the value of u at $z_0 = (x_0, y_0)$. Thus u is constant in D.

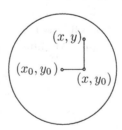

Similarly, v is constant in D. Consequently, $f = u + iv$ is constant in D. ◊

Exercise 2.10. Show that $z \mapsto z^3$ is entire using the Cauchy-Riemann equations.

Exercise 2.11. Show that $z \mapsto \mathrm{Re}(z)$ is complex differentiable nowhere.

Exercise 2.12. Let $D \subset \mathbb{C}$ be a domain. Show, using the Cauchy-Riemann equations, that if $f : D \to \mathbb{C}$ is holomorphic in D, with the property that $f(z) \in \mathbb{R}$ for all $z \in D$, then f is constant in D.

Exercise 2.13. Let $D \subset \mathbb{C}$ be a domain. Show that if $f : D \to \mathbb{C}$ is holomorphic in D, with the property that $f'(z) = 0$ for all $z \in D$, then f is constant in D.

Exercise 2.14. Suppose that $f : \mathbb{C} \to \mathbb{C}$ is an entire function, and let $u := \mathrm{Re}(f)$, $v := \mathrm{Im}(f)$. Suppose, moreover, that there exists a differentiable $h : \mathbb{R} \to \mathbb{R}$ such that $u = h \circ v$. Prove that f must be a constant.

Exercise 2.15. Let $k \in \mathbb{R}$ be a fixed, and let f be defined by $f(z) = (x^2 - y^2) + kxyi$ for $z = x + iy$, $x, y \in \mathbb{R}$. Show that f is entire if and only if $k = 2$.

2.3 Geometric meaning of the complex derivative

In ordinary calculus we have learnt the geometric meaning of the derivative of a real valued function $f : \mathbb{R} \to \mathbb{R}$ at a point $x_0 \in \mathbb{R}$: $f'(x_0)$ is the slope of the tangent to the graph of f at x_0. See Figure 2.8.

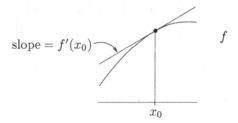

Fig. 2.8 Meaning of $f'(x_0)$.

$$\lim_{x \to x_0} \frac{f(x) - f(x_0)}{x - x_0} = f'(x_0) \text{ implies } \frac{f(x) - f(x_0)}{x - x_0} \approx f'(x_0) \text{ for } x \text{ near } x_0, \text{ i.e.}$$

$$f(x) - f(x_0) \approx f'(x_0)(x - x_0).$$

This means that locally, around x_0, $f(x) - f(x_0)$ looks like the action of the *linear* map $h \mapsto f'(x_0)h : \mathbb{R} \to \mathbb{R}$ on $x - x_0$. Visually this means that near x_0, there is very little difference between the (tangent) line with slope $f'(x_0)$ passing through $(x_0, f(x_0))$ and the graph of f. That is, if we zoom into the graph of the function around the point $(x_0, f(x_0))$, then the graph looks like a straight line.

In this section, we pose an analogous question for a complex valued function map $f : U \to \mathbb{C}$ defined on an open set U, that happens to be complex differentiable at a point z_0:

> What is the geometric meaning of the complex number $f'(z_0)$?

We can't draw a graph of f, because z as well as $f(z)$ belong to $\mathbb{C} = \mathbb{R}^2$, and so $(z, f(z))$ would be a point in $\mathbb{R}^2 \times \mathbb{R}^2 = \mathbb{R}^4$! But we can draw a copy of U in the plane on the left hand side, and on the right hand side, we can imagine a copy of $\mathbb{C}$, with f mapping points from U on the left to points on the right, as shown below.

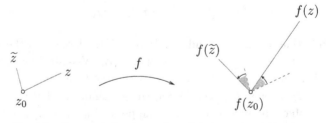

We will now show that the complex number $f'(z_0)$ describes the action of the complex differentiable function locally infinitesimally around z_0 by a anticlockwise rotation through the angle $\mathrm{Arg}(f'(z_0))$ together with a scaling/magnification by $|f'(z_0)|$.

$$\lim_{z \to z_0} \frac{f(z) - f(z_0)}{z - z_0} = f'(z_0), \text{ implies } \frac{f(z) - f(z_0)}{z - z_0} \approx f'(z_0) \text{ for } z \text{ near } z_0, \text{ i.e.}$$

$$f(z) - f(z_0) \approx f'(z_0)(z - z_0).$$

But from the geometric meaning of complex multiplication, we know that when we multiply $z - z_0$ by $f'(z_0)$, $z - z_0$ gets rotated anticlockwise through the angle $\mathrm{Arg}(f'(z_0))$, and the length of $z - z_0$ gets multiplied by the length of $f'(z_0)$, namely we get a magnification by the factor $|f'(z_0)|$. In order to further understand this geometrically, look at Figure 2.9.

Fig. 2.9   Geometric local meaning of the complex derivative.

Suppose that $f'(z_0) = \sqrt{3} + i$, so that $|f'(z_0)| = 2$ and $\mathrm{Arg}(f'(z_0)) = \pi/6$. First look at $z - z_0$ shown in the domain U as the solid line segment between z and z_0. On the right hand side, we have shown a translated version of this line segment as a dashed line, emanating from $f(z_0)$. In order to find out where $f(z)$ is, we just use the fact that $f(z) - f(z_0)$ is approximately equal to $f'(z_0)$ multiplied by $z - z_0$. So the solid line denoting $f(z) - f(z_0)$ on the right hand side is obtained by rotating the rightmost dashed line anticlockwise through an angle of $\mathrm{Arg}(f'(z_0))$ (assumed to be 30° in this picture), and magnifying the length of the dashed line by $|f'(z_0)| = 2$. Note that if want to find the image of another point $\widetilde{z}$ which is near z_0, we have to repeat the same procedure. Namely we first look at the line segment joining z_0 to $\widetilde{z}$, which is the solid line on the left. We have shown a translated version of this as a dashed line, emanating from $f(z_0)$, in the picture on right hand side. In order to find the position of $f(\widetilde{z})$, we first rotate the leftmost dashed line anticlockwise through an angle of $f'(z_0)$, that is 30°, and magnifying the length of the leftmost dashed line by $|f'(z_0)| = 2$. In this manner, we obtain the solid line on the right hand side representing $f(\widetilde{z}) - f(z_0)$. Placing one end at $f(z_0)$, the other end is the point $f(\widetilde{z})$ (approximately!). So we see that locally the action of f is as follows. Imagine the domain as a rubber sheet, and look at a point z_0 on this rubber sheet. Tear out a small portion of this rubber sheet around z_0. The function f takes the point z_0 on this rubber sheet to a point $f(z_0)$ somewhere in the complex plane. If we want to know how the rest of the points on our little torn rubber sheet are mapped by f, one follows this procedure. We place our rubber sheet such that z_0 on our rubber sheet is lying over the point $f(z_0)$ in the complex plane. (Imagine pinning it on the plane with the pin passing through the point marked z_0 on our little torn rubber sheet.) Then we stretch out our rubber sheet about the point z_0 by a factor of $|f'(z_0)|$, and then rotate this stretched rubber sheet anticlockwise by an angle of $\mathrm{Arg}(f'(z_0))$ around the point z_0 on the rubber sheet.

In order to stress this geometric interpretation, let us revisit Example 2.1 yet again, where we considered the squaring map $z \mapsto z^2$. See Figure 2.10 and Example 2.12.

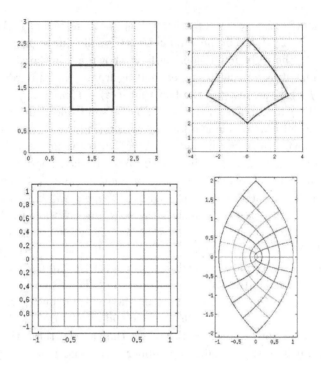

Fig. 2.10   The image of a square and a square grid under the mapping z^2.

Example 2.12. Suppose that we assume complex differentiability of the squaring map f, $z \mapsto z^2$, at a point $z_0 \in \mathbb{C}$. Let us then try to prove that the complex derivative of the squaring map at z_0 must be $2z_0$ by figuring out geometrically the local amplification at z_0 and the local rotation at z_0 produced by the squaring map.

We first ask: what is the local rotation produced? In order to find this out, consider a point z close to z_0 along the ray joining z_0 to z. Look at Figure 2.11, which shows the effect of the squaring map: the angle is doubled, and the distance to 0 is squared. Thus z^2 lies on the ray joining 0 to z_0^2, which makes an angle $2\mathrm{Arg}(z_0)$ with the positive real axis. Hence the line segment joining z_0^2 to z^2 is obtained by rotating the line segment joining z_0 to z anticlockwise through the angle $\mathrm{Arg}(z_0)$. Consequently, $\mathrm{Arg}(f'(z_0)) = \mathrm{Arg}(z_0)$.

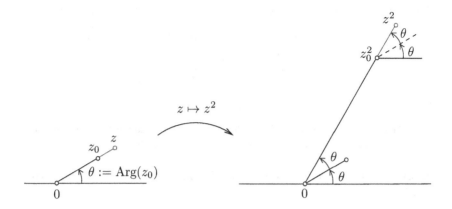

Fig. 2.11 Calculation of the amount of local rotation produced by the squaring map.

Next we ask: what is the local amplification produced? In order to find this out, consider a point z close to z_0 which is at the same distance from 0 as z_0, but it makes an slightly bigger angle $\theta + d\theta$ with the positive real axis. Look at Figure 2.12, which shows the effect of the squaring map. Since $d\theta$ is tiny, the length $|z - z_0|$ is approximately $|z_0| \cdot d\theta$, while the length $|z^2 - z_0^2|$ is approximately $|z_0|^2 \cdot 2d\theta$. Consequently the magnification factor $|f'(z_0)|$ must be equal to $(|z_0|^2 \cdot 2d\theta)/(|z_0| \cdot d\theta) = 2|z_0|$.

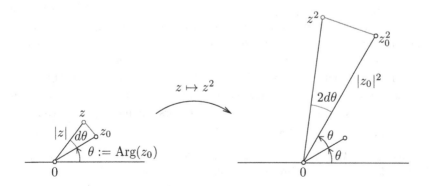

Fig. 2.12 Finding the amount of local magnification produced by the squaring map.

Putting it all together, we have

$$f'(z_0) = |f'(z_0)| \cdot \Big(\cos(\mathrm{Arg}(f'(z_0))) + i \sin(\mathrm{Arg}(f'(z_0))) \Big)$$

$$= 2|z_0| \cdot \Big(\cos(\mathrm{Arg}(z_0)) + i \sin(\mathrm{Arg}(z_0)) \Big)$$

$$= 2z_0.$$

So by investigating the local behaviour of the squaring map f around the point z_0, we could find out the complex derivative $f'(z_0)$. $\Diamond$

Example 2.13. (Complex conjugation is complex differentiable nowhere.) See Figure 2.13.

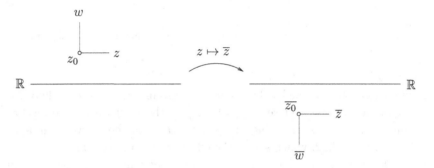

Fig. 2.13 $z \mapsto \overline{z}$ is not holomorphic.

Suppose that $z \mapsto \overline{z}$ is complex differentiable at z_0. Then the local behaviour of the map around z_0 should be a rotation followed by an amplification. Consider the point z near z_0 obtained by a tiny horizontal translation. From the picture, by looking at the images $\overline{z_0}$ and $\overline{z}$, we see that since $\overline{z} - \overline{z_0} = z - z_0$, no rotation is produced. On the other hand, if we look at w which is near z_0 obtained by a tiny vertical displacement, then from the picture, we see that $\overline{w} - \overline{z_0} = -(z - z_0)$, and so there is a rotation through $180°$. But this means that locally the map is not a rotation (because if it were, *all* infinitesimal vectors emanating at z_0 would be rotated by f by a same fixed amount!). $\Diamond$

Exercise 2.16. We know that the power function $z \mapsto z^n$, $n \in \mathbb{N}$, is entire. Find its complex derivative by investigating its local behaviour.
Hint: Since the rotation and magnification is the same for all tiny arrows emanating from z_0, one can simply the argument given in Example 2.12 by looking at what happens to a tiny vector perpendicular to the ray joining 0 to z_0.

Exercise 2.17. We know that the exponential function $z \mapsto e^z$ is entire. Find its complex derivative by investigating its local behaviour.
Hint: Take a point z_0, and move it vertically up through a distance of δ. Look at the image to determine the amplification. Similarly, by moving z_0 horizontally through δ, determine the amount of rotation produced locally.

Exercise 2.18. Give a visual argument to show that the map $z \mapsto \mathrm{Re}(z)$ is not complex differentiable anywhere in $\mathbb{C}$.

Conformality. Look at Figure 1.16 on page 20 again, which shows the action of the entire mapping exp. In this picture, we see that just like in the domain, also the images of the vertical and horizontal lines are mutually perpendicular. We had mentioned that this is the property of conformality, that is, of the preservation of angles between curves in the domain together with their "orientation" is something which is possessed by all complex differentiable functions in domains. Moreover, we had mentioned that this property is possessed by every holomorphic function in its domain. Let us now see why holomorphic functions possess this property, based on what we have learnt about the local action of complex differentiable functions.

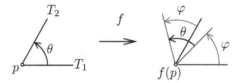

Fig. 2.14   Conformality of a holomorphic f. Here $\varphi = \mathrm{Arg}(f'(p))$.

Let $f : U \to \mathbb{C}$ be holomorphic in U. Imagine two smooth curves intersecting at a point $p \in U$. Since the curves are smooth, they have tangents at p, say T_1 and T_2. Look at Figure 2.14. Near the point p, there is very little difference between the curve and its tangent line at p, so we may assume that the curves are replaced by their tangent lines. These tangent lines make a certain angle. Now let us look at what f does to these lines. Each of the curves is mapped to new curves in $\mathbb{C}$ by f, and these intersect at f of p, that is $f(p)$. They are smooth too, and they possess new tangent lines. But since the local action of f around p is rotation clockwise by $\mathrm{Arg}(f'(p))$ followed by magnification, the new tangent lines are obtained by just rotating clockwise the old tangent lines, and magnifying the image. So it is obvious that the angle will be the same, and so will the orientation. So the conformality of holomorphic maps isn't a mystery anymore!

Relation to real differentiability, and the Cauchy-Riemann equations revisited. Consider a map $f : U \to \mathbb{C}$ which is complex differentiable at $z_0 = (x_0, y_0)$ in the open set U. If u, v denote the real and imaginary part of the function f, then we know that $u, v : U \to \mathbb{R}$ are differentiable at (x_0, y_0). Thus f, that is, the map $(x, y) \mapsto (u(x, y), v(x, y)) : U \to \mathbb{R}^2$, is differentiable (in the real variable sense), and we know that its derivative is the linear transformation

$$A := \begin{bmatrix} \dfrac{\partial u}{\partial x}(x_0, y_0) & \dfrac{\partial u}{\partial y}(x_0, y_0) \\ \dfrac{\partial v}{\partial x}(x_0, y_0) & \dfrac{\partial v}{\partial y}(x_0, y_0) \end{bmatrix}$$

which describes the local action. But since f is complex differentiable, we know that its local action is a anticlockwise rotation by $\theta := \mathrm{Arg}(f'(z_0))$, followed by a magnification by $r := |f'(z_0)|$, and so it is the linear transformation described by the matrix

$$r \begin{bmatrix} \cos\theta & -\sin\theta \\ \sin\theta & \cos\theta \end{bmatrix}.$$

Since A must equal this, we obtain that

$$\frac{\partial u}{\partial x}(x_0, y_0) = r\cos\theta = \frac{\partial v}{\partial y}(x_0, y_0),$$

$$\frac{\partial v}{\partial x}(x_0, y_0) = r\sin\theta = -\frac{\partial u}{\partial y}(x_0, y_0).$$

Moreover, $f'(z_0) = r(\cos\theta + i\sin\theta) = \dfrac{\partial u}{\partial x}(x_0, y_0) + i\dfrac{\partial v}{\partial x}(x_0, y_0)$.

Summarizing, if f is complex differentiable at a point z_0, then it is real differentiable (as a mapping from $U \subset \mathbb{R}^2$ to $\mathbb{R}^2$), but what distinguishes complex differentiability from mere real differentiability is that the real derivative for a complex differentiable mapping is not just any old linear transformation, but a very special one: it is an anticlockwise rotation through an angle θ followed by a scaling by r.

2.4 The d-bar operator

The two Cauchy-Riemann equations can be written as a single equation by introducing what is called the "d-bar operator"

$$\frac{\partial}{\partial \bar{z}}.$$

Let us define the differential operators

$$\frac{\partial}{\partial z} := \frac{1}{2}\left(\frac{\partial}{\partial x} - i\frac{\partial}{\partial y}\right) \text{ and } \frac{\partial}{\partial \bar{z}} := \frac{1}{2}\left(\frac{\partial}{\partial x} + i\frac{\partial}{\partial y}\right).$$

By a differential operator, we mean something which can act on functions and produce new functions. For example, the above two differential operators can act on smooth functions $\varphi : U \to \mathbb{R}$, where U is an open subset of $\mathbb{R}^2$, and by the smoothness of φ, we mean that it possesses at least its first partial derivatives with respect to x and y everywhere in U. Then

$$\frac{\partial \varphi}{\partial z} := \frac{1}{2}\left(\frac{\partial \varphi}{\partial x} - i\frac{\partial \varphi}{\partial y}\right) \text{ and } \frac{\partial \varphi}{\partial \bar{z}} := \frac{1}{2}\left(\frac{\partial \varphi}{\partial x} + i\frac{\partial \varphi}{\partial y}\right).$$

Also, for u, v smooth real-valued functions in U, we define

$$\frac{\partial}{\partial z}(u + iv) := \frac{\partial u}{\partial z} + i\frac{\partial v}{\partial z} \text{ and } \frac{\partial}{\partial \bar{z}}(u + iv) := \frac{\partial u}{\partial \bar{z}} + i\frac{\partial v}{\partial \bar{z}}.$$

With this notation, we have for a holomorphic $f = u + iv$ in the open set $U \subset \mathbb{C}$ that

$$\begin{aligned}
\frac{\partial}{\partial \bar{z}}f &= \frac{\partial}{\partial \bar{z}}(u + iv) = \frac{\partial u}{\partial \bar{z}} + i\frac{\partial v}{\partial \bar{z}} \\
&= \frac{1}{2}\left(\frac{\partial u}{\partial x} + i\frac{\partial u}{\partial y}\right) + i\frac{1}{2}\left(\frac{\partial v}{\partial x} + i\frac{\partial v}{\partial y}\right) \\
&= \frac{1}{2}\left(\frac{\partial u}{\partial x} - \frac{\partial v}{\partial y}\right) + i\frac{1}{2}\left(\frac{\partial u}{\partial y} + \frac{\partial v}{\partial x}\right) \\
&= 0 + i0 = 0,
\end{aligned}$$

using the Cauchy-Riemann equations for the real and imaginary parts u, v of f. Also, we have

$$\begin{aligned}
\frac{\partial}{\partial z}f &= \frac{\partial}{\partial z}(u + iv) = \frac{\partial u}{\partial z} + i\frac{\partial v}{\partial z} \\
&= \frac{1}{2}\left(\frac{\partial u}{\partial x} - i\frac{\partial u}{\partial y}\right) + i\frac{1}{2}\left(\frac{\partial v}{\partial x} - i\frac{\partial v}{\partial y}\right) \\
&= \frac{1}{2}\left(\frac{\partial u}{\partial x} + \frac{\partial v}{\partial y}\right) + i\frac{1}{2}\left(-\frac{\partial u}{\partial y} + \frac{\partial v}{\partial x}\right) \\
&= \frac{1}{2}\cdot 2\frac{\partial u}{\partial x} + i\frac{1}{2}\cdot 2\frac{\partial v}{\partial x} = \frac{\partial u}{\partial x} + i\frac{\partial v}{\partial x} \\
&= f'.
\end{aligned}$$

Summarizing, for f holomorphic in U that $\frac{\partial}{\partial \bar{z}}f = 0$, and $\frac{\partial}{\partial z}f = f'$.

So philosophically, we ought to think of holomorphic functions as "functions of $z, \bar{z}$ which are independent of $\bar{z}$".

Example 2.14. $\bar{z}$ is not holomorphic because

$$\frac{\partial}{\partial \bar{z}} \bar{z} = \frac{\partial}{\partial \bar{z}}(x - iy) = \frac{1}{2}\left(\frac{\partial}{\partial x} + i\frac{\partial}{\partial y}\right)x - i\frac{1}{2}\left(\frac{\partial}{\partial x} + i\frac{\partial}{\partial y}\right)y$$

$$= \frac{1}{2} - i \cdot \frac{1}{2} \cdot i = 1 \neq 0.$$

◇

Example 2.15. $|z|^2 = z\bar{z}$ is not holomorphic in $\mathbb{C} \setminus \{0\}$ because

$$\frac{\partial}{\partial \bar{z}}(z\bar{z}) = \frac{\partial}{\partial \bar{z}}(x^2 + y^2) = \frac{1}{2}\left(\frac{\partial}{\partial x} + i\frac{\partial}{\partial y}\right)(x^2 + y^2)$$

$$= \frac{1}{2}(2x + i2y) = x + iy = z \neq 0 \text{ in } \mathbb{C} \setminus \{0\}.$$

◇

Example 2.16. z^2 is entire because

$$\frac{\partial}{\partial \bar{z}}(z^2) = \frac{\partial}{\partial \bar{z}}(x^2 - y^2 + 2xyi)$$

$$= \frac{1}{2}\left(\frac{\partial}{\partial x} + i\frac{\partial}{\partial y}\right)(x^2 - y^2) + i\frac{1}{2}\left(\frac{\partial}{\partial x} + i\frac{\partial}{\partial y}\right)(2xy)$$

$$= \frac{1}{2}(2x - i2y) + i\frac{1}{2}(2y + i2x) = 0.$$

Moreover,

$$\frac{\partial}{\partial z}(z^2) = \frac{\partial}{\partial z}(x^2 - y^2 + 2xyi)$$

$$= \frac{1}{2}\left(\frac{\partial}{\partial x} - i\frac{\partial}{\partial y}\right)(x^2 - y^2) + i\frac{1}{2}\left(\frac{\partial}{\partial x} - i\frac{\partial}{\partial y}\right)(2xy)$$

$$= \frac{1}{2}(2x + i2y) + i\frac{1}{2}(2y - i2x) = 2(x + iy) = 2z.$$

◇

Exercise 2.19. Show that $4\dfrac{\partial}{\partial z}\dfrac{\partial}{\partial \bar{z}} = \Delta$, where $\Delta := \dfrac{\partial^2}{\partial x^2} + \dfrac{\partial^2}{\partial y^2}$ is the Laplacian.

2.5 Notes

The section on the geometric meaning of the complex derivative follows closely the exposition in [Needham (1997)]. Exercises 2.16 and 2.17 are taken from [Needham (1997)].

Chapter 3

Cauchy Integral Theorem and consequences

Having become familiar with complex differentiation, we now turn to integration. In this chapter we will learn a very important theorem in complex analysis, called

> The Cauchy Integral Theorem.

We will begin by defining "contour integration". And then we will show the Cauchy Integral Theorem. One might ask: Why is this contour integration and the Cauchy Integral Theorem so important? The importance of integration in the complex plane stems from the fact that it will lead to a greater understanding of holomorphic functions, for example, the fundamental fact that holomorphic functions are infinitely many times complex differentiable! In this chapter we will learn the following main topics:

(1) Contour integration and its properties.
(2) The Fundamental Theorem of Contour Integration.
(3) The Cauchy Integral Theorem.
(4) Consequences of the Cauchy Integral Theorem:

    (a) Existence of a primitive.
    (b) Infinite differentiability of holomorphic functions.
    (c) Liouville's Theorem and the Fundamental Theorem of Algebra.
    (d) Morera's Theorem.

3.1 Definition of the contour integral

In ordinary calculus, given a continuous function $f : [a, b] \to \mathbb{R}$,

$$\int_a^b f(x) dx \tag{3.1}$$

has a clear meaning. Now suppose we wish to generalize this in the complex setting: given z, w complex numbers, want to give meaning to something like

$$\int_z^w f(\zeta)\,d\zeta.$$

Then a first question is: How do we get from z to w?

In $\mathbb{R}$, if $a < b$, then there is just one way of going from the real number a to the real number b, and so our data in the real case is just:

(1) $a < b$, and
(2) a continuous function $f : [a, b] \to \mathbb{R}$.

But now z and w are points in the complex plane, and so there are many possible connecting paths along which we could integrate. See Figure 3.1.

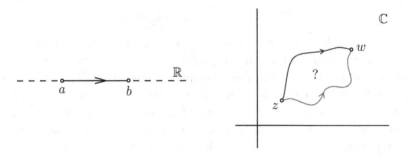

Fig. 3.1   Which path to go from z to w?

So in the complex setting, besides specifying the end points z and w, we will also specify the path γ taken to go from z to w, and we will replace the above expression (3.1) in the real case by an expression which looks like this in the complex setting:

$$\int_\gamma f(z)\,dz.$$

We call such an expression a "contour" integral, for the computation of which we need the following data:

(1) A domain D $(\subset \mathbb{C})$, and $z, w \in D$.
(2) A continuous function $f : D \to \mathbb{C}$.
(3) A **smooth** path $\gamma : [a, b] \to D$ joining z to w.

We note that we need not merely a path, but a **smooth** path joining z to w. What do we mean by "smooth" here? Recall that a path $\gamma : [a, b] \to D$ is a continuous function. We can decompose γ into its real and imaginary parts:

$$\gamma(t) = x(t) + iy(t), \quad t \in [a, b],$$

where $x, y : [a, b] \to \mathbb{R}$. The path γ is called *smooth* if x, y are continuously differentiable. Here are some examples.

Example 3.1. Let $\gamma : [0, 1] \to \mathbb{C}$ be given by $\gamma(t) = t(1+i)$, $t \in [0, 1]$. Then the real and imaginary parts $x, y : [0, 1] \to \mathbb{R}$ of γ are given by $x(t) = t$, $y(t) = t$, $t \in [0, 1]$. Since x, y are continuously differentiable on $[0, 1]$, γ is a smooth path. See Figure 3.2.

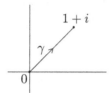

Fig. 3.2   The smooth path γ.

Similarly consider the two paths $\gamma_1, \gamma_2 : [0, 2\pi] \to \mathbb{R}$, given by

$$\gamma_1(t) = \exp(it) \quad \text{and} \quad \gamma_2(t) = \exp(2it),$$

where $t \in [0, 2\pi]$. Then since the real and imaginary parts of these paths are $\cos t, \sin t, \cos(2t), \sin(2t)$, all of which are continuously differentiable, it follows that γ_1, γ_2 are smooth paths. See Figure 3.3.

Fig. 3.3   The smooth path γ.

So the set of image points (the ranges of γ_1 and γ_2) is the same, that is: $\{\gamma_1(t) : t \in [0, 2\pi]\} = \{\gamma_2(t) : t \in [0, 2\pi]\} = \{z \in \mathbb{C} : |z| = 1\}$, (the unit circle with center 0). However γ_1 and γ_2 are *different paths*, because the functions are not the same: for example, $\gamma_1(\pi) = -1 \neq 1 = \gamma_2(\pi)$. $\Diamond$

Remark 3.1. It is very common and convenient to refer to the *range*

$$\{\gamma(t) : t \in [a,b]\}$$

of a path $\gamma : [a,b] \to \mathbb{C}$ as the *path/curve* itself. With this usage, a path becomes a concrete geometric object (as opposed to being a mapping), such as a circle or a straight line segment in the complex plane and hence can be easily visualized. The difficulty with this abuse of terminology is that several *different* paths can have the same image, and so it causes ambiguity.

The precise definition of the contour integral is given below.

Definition 3.1. Given

(1) a domain D,
(2) a continuous function $f : D \to \mathbb{C}$ (with real and imaginary parts denoted by $u, v : D \to \mathbb{R}$), and
(3) a smooth path $\gamma : [a,b] \to D$ (with real and imaginary parts denoted by $x, y : [a,b] \to \mathbb{R}$),

we define

$$\int_\gamma f(z)dz := \int_a^b f(\gamma(t))\gamma'(t)dt \tag{3.2}$$

$$:= \int_a^b \Big(u(\gamma(t)) + iv(\gamma(t)) \Big) \cdot (x'(t) + iy'(t))dt$$

$$:= \int_a^b \Big(u(\gamma(t)) \cdot x'(t) - v(\gamma(t)) \cdot y'(t) \Big) dt$$

$$+ i\int_a^b \Big(v(\gamma(t)) \cdot x'(t) + u(\gamma(t)) \cdot y'(t) \Big) dt.$$

We note that the two integrals in the last line above are just the usual Riemann integrals of real-valued continuous functions.

The contour integral can be interpreted geometrically as follows. The term

$$\gamma'(t)dt = x'(t)dt + iy'(t)dt$$

can be viewed as an infinitesimal incremental piece of the contour. We multiply this by the (almost constant) value $f(\gamma(t))$ of f on this incremental piece. Finally, we add up all these contributions along the contour to get the total as the integral

$$\int_a^b f(\gamma(t))\gamma'(t)dt.$$

See Figure 3.4.

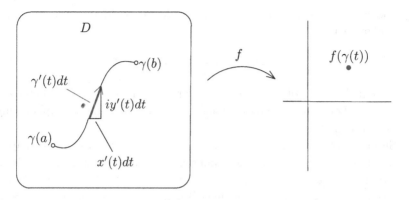

Fig. 3.4   Geometric meaning of the contour integral.

Here is an example.

Example 3.2. Let

(1) $D = \mathbb{C}$,
(2) γ be the smooth path given by $\gamma(t) = t(1 + i)$, $t \in [0, 1]$, and
(3) $f = (z \mapsto \bar{z})$.

Then

$$\int_\gamma f(z)dz = \int_0^1 \overline{t(1 + i)} \cdot (1 + i)dt$$

$$= \int_0^1 t(1 - i) \cdot (1 + i)dt = \int_0^1 t(1^2 - i^2)dt = \int_0^1 t(1 + 1)dt$$

$$= 2\int_0^1 tdt = 2 \cdot \frac{t^2}{2}\Big|_0^1 = 2 \cdot \frac{1}{2} = 1.$$

$\Diamond$

Exercise 3.1. Consider the three paths $\gamma_1, \gamma_2, \gamma_3 : [0, 2\pi] \to \mathbb{C}$ defined by

$$\gamma_1(t) = \exp(it),$$
$$\gamma_2(t) = \exp(2it),$$
$$\gamma_3(t) = \exp(-it),$$

for $t \in [0, 2\pi]$. Show that their images are the same, but the three contour integrals

$$\int_{\gamma_1} \frac{1}{z}dz, \quad \int_{\gamma_2} \frac{1}{z}dz, \quad \int_{\gamma_3} \frac{1}{z}dz$$

are all different.

Exercise 3.2. Let f be holomorphic in a domain and let $\gamma : [0,1] \to D$ be a smooth path. Show that

$$\frac{d}{dt}f(\gamma(t)) = f'(\gamma(t)) \cdot \gamma'(t) \text{ for all } t \in [0,1].$$

In the sequel, we will often assume that our smooth paths are parameterized by $[0,1]$ (rather than some more general interval $[a,b]$). Let us explain why we may assume this.

Suppose that $\gamma : [a,b] \to \mathbb{C}$ and $\tilde{\gamma} : [c,d] \to \mathbb{C}$, are two smooth paths, such that there is a continuously differentiable function $\varphi : [c,d] \to [a,b]$ such that $a = \varphi(c)$, $b = \varphi(d)$, and $\tilde{\gamma}(t) = \gamma(\varphi(t))$ for $t \in [a,b]$. We call such smooth paths "equivalent". (Imagine going from $\gamma(a) = \tilde{\gamma}(c)$ to $\gamma(b) = \tilde{\gamma}(d)$ along the same route, but with possibly different speeds.) See Figure 3.5. We now show the following.

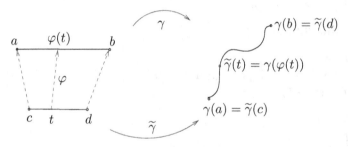

Fig. 3.5   Equivalent paths.

Equivalent paths give the same integral. By the chain rule it follows that

$$\int_{\tilde{\gamma}} f(z)dz \quad = \quad \int_a^b f(\tilde{\gamma}(t))\tilde{\gamma}'(t)dt = \int_a^b f(\gamma(\varphi(t)))\gamma'(\varphi(t))\varphi'(t)dt$$

$$\overset{(\tau = \varphi(t))}{=} \int_c^d f(\gamma(\tau))\gamma'(\tau)d\tau = \int_\gamma f(z)dz.$$

In particular, given any $\gamma : [a,b] \to \mathbb{C}$, we can define $\varphi : [0,1] \to [a,b]$ by

$$\varphi(t) = (1-t)a + tb, \quad t \in [a,b].$$

Then φ is continuously differentiable, and $\varphi(0) = a$, $\varphi(1) = b$. So with $c := 0$, $d := 1$ in the above, and with $\tilde{\gamma} : [0,1] \to \mathbb{C}$ defined by $\tilde{\gamma} = \gamma \circ \varphi$, we see that

$$\int_{\tilde{\gamma}} f(z)dz = \int_\gamma f(z)dz,$$

and so there is no loss of generality (when it comes to statements about contour integrals) in assuming that the smooth path is parameterized by $[0,1]$.

Contour integrals along piecewise smooth paths. We extend the definition above to paths with "corners". A path $\gamma : [a,b] \to \mathbb{C}$ is called a *piecewise smooth path/curve* if there exist points $c_1,\ldots,c_n$ such that

$$a < c_1 < \cdots < c_n < b$$

and γ is continuously differentiable on $[a,c_1], [c_1,c_2], \ldots, [c_{n-1},c_n], [c_n,b]$. For such a path, we define

$$\int_\gamma f(z)dz := \int_a^{c_1} f(\gamma(t))\gamma'(t)dt + \int_{c_1}^{c_2} f(\gamma(t))\gamma'(t)dt + \cdots$$

$$+ \int_{c_{n-1}}^{c_n} f(\gamma(t))\gamma'(t)dt + \int_{c_n}^{b} f(\gamma(t))\gamma'(t)dt.$$

Here is an example.

Example 3.3. Let $\widetilde{\gamma}$ be the path from 0 to $1+i$ given by

$$\widetilde{\gamma}(t) = \begin{cases} t & \text{if } t \in [0,1] \\ 1 + (t-1)i & \text{if } t \in (1,2]. \end{cases}$$

See Figure 3.6.

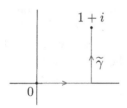

Fig. 3.6 The piecewise smooth path $\widetilde{\gamma}$.

Then we have

$$\int_{\widetilde{\gamma}} \overline{z}\,dz = \int_0^1 \overline{t}1dt + \int_1^2 \overline{(1+(t-1)i)}idt = \int_0^1 tdt + \int_1^2 (1-(t-1)i)idt$$

$$= \int_0^1 tdt + \int_1^2 (i + (t-1))dt$$

$$= \frac{1}{2} + i + \frac{4-1}{2} - 1 = 1 + i.$$

$\diamond$

Let us look at the answers obtained in Examples 3.2 and 3.3. We see that the integrand is the same (the nonholomorphic function $z \mapsto \bar{z}$), the end points of the paths are the same (namely 0 and $1+i$), and the paths joining these two points are different (γ and $\tilde{\gamma}$). See Figure 3.7.

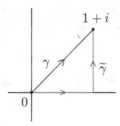

Fig. 3.7   The two paths γ and $\tilde{\gamma}$.

The answer we obtain in each case is different too:

$$\int_{\gamma} \bar{z}\,dz = 1 \neq 1+i = \int_{\tilde{\gamma}} \bar{z}\,dz.$$

Thus the integral *depends* on the path for the nonholomorphic integrand $z \mapsto \bar{z}$. This is not strange, because from the definition of the contour integral of course we expect the value of the contour integral to depend on the route chosen. The main goal in this chapter is to show that the contour integration of a holomorphic function along two paths joining the same points from z to w is the *same* provided that the map is holomorphic everywhere in the region between the two paths! It turns out that this is a fundamental result (called the Cauchy Integral Theorem) in complex analysis because many further results follow from this. Let us check this in an example.

Example 3.4. Consider the same two contours γ, $\tilde{\gamma}$ considered in Examples 3.2 and 3.3 above. But instead of taking the *nonholomorphic* map $z \mapsto \bar{z}$, consider now the *entire* function z: Then we have

$$\int_{\gamma} z\,dz = \int_0^1 (1+i)t(1+i)dt = \int_0^1 2it\,dt = i, \text{ and}$$

$$\int_{\tilde{\gamma}} z\,dz = \int_0^1 t \cdot 1\,dt + \int_1^2 (1+(t-1)i)i\,dt$$

$$= \int_0^1 t\,dt + \int_1^2 (i-(t-1))dt = \frac{1}{2} - \frac{1}{2} + i = i.$$

We note that this time the answer is the same for γ and for $\tilde{\gamma}$.   ◊

Exercise 3.3. Integrate the following functions over the circle $|z| = 2$, oriented anticlockwise:

(1) $z + \bar{z}$.

(2) $z^2 - 2z + 3$.

(3) xy, where $z = x + iy$, $x, y \in \mathbb{R}$.

Exercise 3.4. Evaluate $\displaystyle\int_{\gamma} \mathrm{Re}(z)dz$, where γ is:

(1) The straight line segment from 0 to $1 + i$.

(2) The short circular arc with center i and radius 1 joining 0 to $1 + i$.

(3) The part of the parabola $y = x^2$ from $x = 0$ to $x = 1$ joining 0 to $1 + i$.

3.1.1 *An important integral*

Let us now calculate a very important simple contour integral, which will recur again and again in the course. We fix a useful convention: throughout the book, unless otherwise stated, a *circular path with center z_0 and radius $r > 0$ traversed in the anticlockwise direction* will mean the path $C : [0, 2\pi] \to \mathbb{C}$ given by $C(t) = z_0 + r\exp(it)$, $t \in [0, 2\pi]$. (Thus the circle is traversed *once*.) See Figure 3.8.

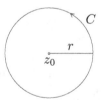

Fig. 3.8 The circular path C with center z_0 and radius r traversed in the anticlockwise direction.

We will now calculate the integrals $\displaystyle\int_{C} (z - z_0)^n dz$, $n \in \mathbb{Z}$.
This calculation will prove to be very useful later on.

Theorem 3.1. *Let C be a circular path with center z_0 and radius $r > 0$ traversed in the anticlockwise direction. Then*

$$\int_{C} (z - z_0)^n dz = \begin{cases} 2\pi i & \text{if } n = -1, \\ 0 & \text{if } n \neq -1. \end{cases}$$

We note that the answer is independent of r.

Proof.   We have $C(t) = z_0 + r\exp(it) = z_0 + r\cos t + ir\sin t$, $t \in [0, 2\pi]$, and so $C'(t) = -r\sin t + ir\cos t = ir(\cos t + i\sin t) = ir\exp(it)$, $t \in [0, 2\pi]$. We now consider the two cases:

1° When $n = -1$, we have

$$\int_C (z - z_0)^n dz = \int_C (z - z_0)^{-1} dz = \int_0^{2\pi} \frac{1}{r\exp(it)} \cdot ir\exp(it)dt$$

$$= \int_0^{2\pi} i\, dt = 2\pi i.$$

2° When $n \neq -1$, we have

$$\int_C (z - z_0)^n dz = \int_0^{2\pi} r^n \exp(nit) \cdot ir\exp(it)dt$$

$$= \int_0^{2\pi} ir^{n+1} \exp(i(n+1)t)dt$$

$$= -r^{n+1} \int_0^{2\pi} \sin((n+1)t)dt + ir^{n+1} \int_0^{2\pi} \cos((n+1)t)dt$$

$$= 0 + 0 = 0.$$

This completes the proof. $\square$

We will see later that this has significant consequences. For instance, suppose that we have an f that has a "expansion in terms of integral powers of z" (whatever that means), in an annulus $\mathbb{A} := \{z \in \mathbb{C} : r < |z - z_0| < R\}$ with center z_0, inner radius r, and outer radius R like this:

$$f(z) = \sum_{n \in \mathbb{Z}} a_n(z - z_0)^n, \quad z \in \mathbb{A}.$$

We will give precise meaning to the (infinite) "sum" above later on, but for now, one may just imagine a finite sum (so that all but finitely many a_ns are zeros). Then by multiplying both sides by $(z - z_0)^{-(m+1)}$ for some $m \in \mathbb{Z}$, we obtain

$$\frac{f(z)}{(z - z_0)^{m+1}} = \sum_{n \in \mathbb{Z}} a_n(z - z_0)^{n-m-1},$$

and so

$$\frac{1}{2\pi i} \int_C \frac{f(z)}{(z - z_0)^{m+1}} dz = \sum_{n \in \mathbb{Z}} a_n \int_C (z - z_0)^{n-m-1} dz = a_m.$$

In the above, we assumed that the sum passes through integration over C, which for finite sums, follows from the definition of the integral, and we will see this in the next section. When the sum is not finite, we will make precise the details later, but things work out essentially as suggested by this calculation. The upshot of it all is that the coefficients are expressible in terms of a contour integral, and we will see later that any function holomorphic in an annulus will have such an expansion.

Exercise 3.5. Let C be the circular path with center 0 and radius 1 traversed in the anticlockwise direction. Show that for $0 \leq k \leq n$,

$$\binom{n}{k} = \frac{1}{2\pi i} \int_C \frac{(1+z)^n}{z^{k+1}} dz.$$

3.2 Properties of contour integration

In this section we will show some useful properties of contour integration. The following result follows in a straightforward manner from the definition of contour integration.

Proposition 3.1. *Let D be a domain in $\mathbb{C}$ and $\gamma : [a,b] \to D$ be a piecewise smooth path. Then the following hold:*

(1) *For all continuous $f, g : D \to \mathbb{C}$,*

$$\int_\gamma (f+g)(z)\,dz = \int_\gamma f(z)\,dz + \int_\gamma g(z)\,dz.$$

(2) *For all continuous $f : D \to \mathbb{C}$ and all $\alpha \in \mathbb{C}$,*

$$\int_\gamma (\alpha f)(z)\,dz = \alpha \int_\gamma f(z)\,dz.$$

Let $C(D; \mathbb{C})$ denote the vector space over $\mathbb{C}$ of all complex-valued continuous functions on D with pointwise operations. Then the above result says that each piecewise smooth path γ in D induces a linear transformation from $C(D; \mathbb{C})$ to $\mathbb{C}$, namely

$$f \mapsto \int_\gamma f(z)\,dz : C(D; \mathbb{C}) \to \mathbb{C}.$$

Exercise 3.6. Prove Proposition 3.1.

Opposite paths. Given a smooth path $\gamma : [a, b] \to D$ in a domain D, its *opposite path*, $-\gamma : [a, b] \to D$, is defined by $(-\gamma)(t) = \gamma(a+b-t)$, $t \in [a, b]$. Then $(-\gamma)(a) = \gamma(b)$ and $(-\gamma)(b) = \gamma(a)$, and so $-\gamma$ starts where γ ends, and ends at the starting point of γ while traversing the same path of γ but in the opposite direction. See Figure 3.9.

Fig. 3.9   The opposite path $-\gamma$ to the path γ.

But why do we denote the opposite path by $-\gamma$? Here's why.

Proposition 3.2. *Let* $\gamma : [a, b] \to D$ *be a smooth path in a domain D and* $f : D \to \mathbb{C}$ *be a continuous function. Then*

$$\int_{-\gamma} f(z)dz = -\int_{\gamma} f(z)dz.$$

Proof. We have

$$
\begin{aligned}
\int_{-\gamma} f(z)dz &= \int_a^b f((-\gamma)(t)) \cdot (-\gamma)'(t)dt \\
&= \int_a^b f(\gamma(a + b - t)) \cdot (\gamma'(a + b - t)) \cdot (-1)dt \\
&\overset{(\tau=a+b-t)}{=} \int_b^a f(\gamma(\tau)) \cdot \gamma'(\tau)d\tau = -\int_a^b f(\gamma(\tau)) \cdot \gamma'(\tau)d\tau \\
&= -\int_{\gamma} f(z)dz.
\end{aligned}
$$

$\square$

Exercise 3.7. Show that $-(-\gamma) = \gamma$, where $\gamma : [a, b] \to D$ is a smooth path in a domain D.

Concatenation of paths. Let D be a domain and let

$$\gamma_1 : [a_1, b_1] \to D \text{ and}$$
$$\gamma_2 : [a_2, b_2] \to D$$

be two paths such that

$$\gamma_1(b_1) = \gamma_2(a_2)$$

(so that γ_2 starts where γ_1 ends). Define $\gamma_1 + \gamma_2 : [a_1, b_1 + b_2 - a_2]$ to be their "concatenation" by:

$$(\gamma_1 + \gamma_2)(t) = \begin{cases} \gamma_1(t) & \text{for } a_1 \le t \le b_1, \\ \gamma_2(t - b_1 + a_2) & \text{for } b_1 \le t \le b_1 + b_2 - a_2. \end{cases}$$

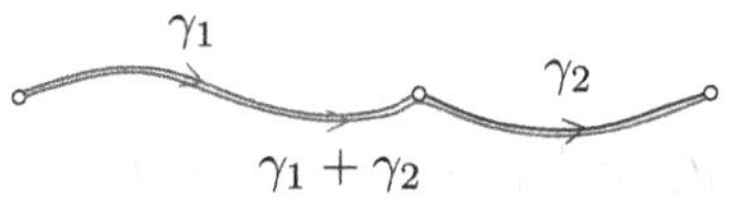

Fig. 3.10   Concatenation $\gamma_1 + \gamma_2$ of two paths γ_1, γ_2.

Proposition 3.3. *Let D be a domain in $\mathbb{C}$ and let $\gamma_1 : [a_1, b_1] \to D$ and $\gamma_2 : [a_2, b_2] \to D$ be two paths such that $\gamma_1(b_1) = \gamma_2(a_2)$. Then*

$$\int_{\gamma_1 + \gamma_2} f(z)\, dz = \int_{\gamma_1} f(z)\, dz + \int_{\gamma_2} f(z)\, dz.$$

Proof. We have

$$\int_{\gamma_1 + \gamma_2} f(z)\, dz = \int_{a_1}^{b_1 + b_2 - a_2} f((\gamma_1 + \gamma_2)(t))(\gamma_1 + \gamma_2)'(t)dt$$

$$= \int_{a_1}^{b_1} f((\gamma_1 + \gamma_2)(t))(\gamma_1 + \gamma_2)'(t)dt$$
$$+ \int_{b_1}^{b_1 + b_2 - a_2} f((\gamma_1 + \gamma_2)(t))(\gamma_1 + \gamma_2)'(t)dt$$

$$= \int_{a_1}^{b_1} f(\gamma_1(t))\gamma_1'(t)dt$$
$$+ \int_{b_1}^{b_1 + b_2 - a_2} f(\gamma_2(\tau - b_1 + a_2))\gamma_2'(\tau - b_1 + a_2)d\tau$$

$$= \int_{\gamma_1} f(z)\, dz + \int_{a_2}^{b_2} f(\gamma_2(s))\gamma_2'(s)dt \quad (s = \tau - b_1 + a_2)$$

$$= \int_{\gamma_1} f(z)\, dz + \int_{\gamma_2} f(z)\, dz. \qquad \square$$

Exercise 3.8. If $\gamma : [a, b] \to D$ is a smooth path in a domain D, and $f : D \to \mathbb{C}$ is continuous, then show that $\displaystyle \int_{\gamma + (-\gamma)} f(z)dz = 0$.

A useful estimate. We now prove an inequality for the size of the contour integral in terms of the size of $|f|$ along the contour, and the length of the contour. This will prove to be indispensable in the sequel.

Proposition 3.4. *Let*

(1) *D be a domain in $\mathbb{C}$,*
(2) *$\gamma : [a, b] \to D$ be a piecewise smooth path and*
(3) *$f : D \to \mathbb{C}$ be a continuous function.*

Then

$$\left| \int_\gamma f(z)\, dz \right| \leq \left(\max_{t \in [a,b]} |f(\gamma(t))| \right) \cdot (\text{length of } \gamma). \qquad (3.3)$$

Recall that the length of γ is given by

$$\int_a^b \sqrt{(x'(t))^2 + (y'(t))^2},$$

where $x, y : [a, b] \to \mathbb{R}$ are the real and imaginary parts of γ. See Figure 3.11.

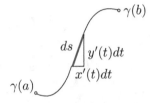

Fig. 3.11   The arc length of the path γ is the sum of the incremental arc lengths ds, where $ds = \sqrt{(x'(t)dt)^2 + (y'(t)dt)^2} = \sqrt{(x'(t))^2 + (y'(t))^2}\, dt$.

Proof.   Consider first a curve $\varphi : [a, b] \to \mathbb{C}$, for which we prove

$$\left| \int_a^b \varphi(t)\, dt \right| \leq \int_a^b |\varphi(t)|\, dt.$$

To see this, let $\int_a^b \varphi(t)\, dt = r \cdot \exp(i\theta)$, where $r \geq 0$ and $\theta \in (-\pi, \pi]$. Then

$$\left| \int_a^b \varphi(t)\, dt \right| = r = \exp(-i\theta) \cdot r \cdot \exp(i\theta)$$

$$= \exp(-i\theta) \cdot \int_a^b \varphi(t)\, dt = \int_a^b \exp(-i\theta) \cdot \varphi(t)\, dt$$

$$= \int_a^b \text{Re}\left(\exp(-i\theta) \cdot \varphi(t) \right) dt + i \int_a^b \text{Im}\left(\exp(-i\theta) \cdot \varphi(t) \right) dt.$$

But the left hand side is real, and so the integral of the imaginary part on the right hand side must be zero. Consequently,

$$\left| \int_a^b \varphi(t)dt \right| = \int_a^b \operatorname{Re}\left(\exp(-i\theta) \cdot \varphi(t)\right) dt$$

$$\leq \int_a^b |\operatorname{Re}\left(\exp(-i\theta) \cdot \varphi(t)\right)| \, dt$$

$$\leq \int_a^b |\exp(-i\theta) \cdot \varphi(t)| \, dt = \int_a^b |\varphi(t)| \, dt.$$

The proposition now follows, since with $\varphi(t) := f(\gamma(t)) \cdot \gamma'(t)$, $t \in [a, b]$,

$$\left| \int_\gamma f(z) \, dz \right| = \left| \int_a^b f(\gamma(t))\gamma'(t)dt \right|$$

$$\leq \int_a^b |f(\gamma(t))\gamma'(t)|dt = \int_a^b |f(\gamma(t))||\gamma'(t)|dt$$

$$\leq \left(\max_{t \in [a,b]} |f(\gamma(t))| \right) \int_a^b |\gamma'(t)|dt.$$

If $\gamma(t) = x(t) + iy(t)$, where x, y are real-valued, then

$$\int_a^b |\gamma'(t)|dt = \int_a^b \sqrt{(x'(t))^2 + (y'(t))^2}dt = \text{length of } \gamma.$$

This completes the proof. $\qquad\square$

Exercise 3.9. Calculate the upper bound given by (3.3) on the absolute value of the integral

$$\int_\gamma z^2 \, dz,$$

where γ is the straight line path from 0 to $1 + i$. Also, compute the integral and find its absolute value.

Exercise 3.10. Using the calculation done in Exercise 3.5, deduce that $\binom{2n}{n} \leq 4^n$.

3.3 Fundamental Theorem of Contour Integration

Let us recall the Fundamental Theorem of Calculus in the real setting:

Theorem 3.2. (Fundamental Theorem of Calculus) *If $F : [a, b] \to \mathbb{R}$ is continuously differentiable and $F' =: f$ on $[a, b]$, then*

$$\int_a^b f(x)dx = F(b) - F(a).$$

This is an important result, because it facilitates the computation of the Riemann integral. Indeed, if we know that a function is the derivative of something, then it is easy to calculate its integral. For example,

$$\int_a^b x^2 dx = \frac{b^3 - a^3}{3} \quad \text{because} \quad x^2 = \frac{d}{dx}\left(\frac{x^3}{3}\right).$$

Analogously, we will now see that if f is the derivative of a holomorphic function, then the calculation of the contour integral

$$\int_\gamma f(z)\,dz$$

is easy, since we have (similar to the Fundamental Theorem of Calculus in the real setting):

Theorem 3.3. (Fundamental[1] Theorem of Contour Integration) *Let*

(1) D *be a domain in* $\mathbb{C}$,
(2) $\gamma : [a, b] \to D$ *be a piecewise smooth path,*
(3) $f : D \to \mathbb{C}$ *be a continuous function in* D,
(4) $F : D \to \mathbb{C}$ *be a holomorphic function such that* $F' = f$ *in* D.

Then $\int_\gamma f(z)\,dz = F(\gamma(b)) - F(\gamma(a))$.

How does this theorem help? One can now calculate some contour integrals very easily (just like in ordinary calculus). Here is an example.

Example 3.5. Since $\dfrac{d}{dz}\left(\dfrac{z^2}{2}\right) = z$ $(z \in \mathbb{C})$, for *any* γ joining 0 to $1 + i$,

$$\int_\gamma z\,dz = \frac{(1+i)^2}{2} - \frac{0^2}{2} = \frac{1 + 2i + i^2}{2} = \frac{1 + 2i - 1}{2} = i,$$

and so in particular, we recover the answer obtained in Example 3.4. $\Diamond$

In particular, as we have also seen in the previous example,

$$\int_\gamma f(z)\,dz = F(w) - F(z)$$

is independent of the path γ joining the points z to w, when f possesses an "antiderivative" or "primitive" F in D.

---

[1]The naming of this result is done just to highlight the similarity with the real analysis analogue. However, in complex analysis, this isn't all that "fundamental". We will soon learn about Cauchy's Integral Theorem, which is certainly more fundamental!

Example 3.6. There is no function $F : \mathbb{C} \to \mathbb{C}$ such that $F'(z) = \bar{z}$ for all $z \in \mathbb{C}$. Indeed, the calculation in Examples 3.2 and 3.3 shows that the contour integral along paths joining 0 to $1 + i$ does depend on the path chosen. ◊

Proof. (of Theorem 3.3.) For $z = x + iy \in D$, where x, y are real, define the real-valued functions U, V, u, v by

$$F(x + iy) = U(x, y) + iV(x, y),$$
$$f(x + iy) = u(x, y) + iv(x, y).$$

Also, set $\gamma(t) = x(t) + iy(t)$ ($t \in [a, b]$), where x, y are real-valued. Then by the Cauchy-Riemann equations, we have

$$u(x, y) + iv(x, y) = f(x + iy) = F'(x + iy)$$

$$= \frac{\partial U}{\partial x}(x, y) + i\frac{\partial V}{\partial x}(x, y) = \frac{\partial V}{\partial y}(x, y) - i\frac{\partial U}{\partial y}(x, y).$$

By the chain rule and the above, we have

$$\frac{d}{dt}U(x(t), y(t)) = \frac{\partial U}{\partial x}(x(t), y(t)) \cdot x'(t) + \frac{\partial U}{\partial y}(x(t), y(t)) \cdot y'(t)$$

$$= u(x(t), y(t)) \cdot x'(t) - v(x(t), y(t)) \cdot y'(t).$$

Similarly,

$$\frac{d}{dt}V(x(t), y(t)) = \frac{\partial V}{\partial x}(x(t), y(t)) \cdot x'(t) + \frac{\partial V}{\partial y}(x(t), y(t)) \cdot y'(t)$$

$$= v(x(t), y(t)) \cdot x'(t) + u(x(t), y(t)) \cdot y'(t).$$

Thus

$$\int_\gamma f(z)\,dz = \int_a^b f(\gamma(t))\gamma'(t)dt$$

$$= \int_a^b \Big(u(x(t), y(t)) + iv(x(t), y(t)) \Big)(x'(t) + iy'(t))dt$$

$$= \int_a^b \frac{d}{dt}U(x(t), y(t))dt + i\int_a^b \frac{d}{dt}V(x(t), y(t))dt$$

$$= U(x(b), y(b)) - U(x(a), y(a)) + i(V(x(b), y(b)) - V(x(a), y(a)))$$

$$= F(\gamma(b)) - F(\gamma(a)).$$

This completes the proof. □

Exercise 3.11. Show, using the Cauchy-Riemann equations, that $z \mapsto \bar{z}$ has no primitive in $\mathbb{C}$.

Exercise 3.12. (Integration by Parts Formula.) Let f, g be holomorphic functions defined in a domain D, such that f', g' are continuous in D, and let γ be a piecewise smooth path in D from $w \in D$ to $z \in D$. Show that

$$\int_\gamma f(\zeta)g'(\zeta)d\zeta = f(z)g(z) - f(w)g(w) - \int_\gamma f'(\zeta)g(\zeta)d\zeta.$$

Exercise 3.13. Evaluate $\int_\gamma \cos z \, dz$, where γ is any path joining $-i$ to i.

Definition 3.2. A path $\gamma : [a, b] \to \mathbb{C}$ is said to be *closed* if $\gamma(a) = \gamma(b)$.

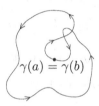

Corollary 3.1. *Let*

(1) *D be a domain in $\mathbb{C}$,*
(2) *$\gamma : [a, b] \to D$ be a closed piecewise smooth path,*
(3) *$f : D \to \mathbb{C}$ is a continuous function in D,*
(4) *$F : D \to \mathbb{C}$ be a holomorphic function such that $F' = f$ in D.*

Then $\int_\gamma f(z) \, dz = 0$.

Proof. $\int_\gamma f(z)dz = F(\gamma(b)) - F(\gamma(a)) = 0$ since $\gamma(b) = \gamma(a)$. $\square$

Example 3.7. For $m \in \mathbb{Z} \setminus \{0\}$, $z \in D := \mathbb{C} \setminus \{0\}$, $\dfrac{d}{dz}\left(\dfrac{z^m}{m}\right) = z^{m-1}$; so

$$\int_\gamma z^{m-1} dz = 0$$

for any closed path γ in D. What if $m = 0$? Note that $\mathrm{Log}' z = 1/z$ for $z \in \widetilde{D} := \mathbb{C} \setminus (-\infty, 0)$, and so for any path $\widetilde{\gamma}$ in $\widetilde{D}$, we do have

$$\int_{\widetilde{\gamma}} \frac{1}{z} \, dz = 0.$$

However, in D, $\dfrac{1}{z}$ doesn't have a primitive; see Exercise 3.16. $\Diamond$

Exercise 3.14. Use the Fundamental Theorem of Contour Integration to write down the value of
$$\int_{\gamma} \exp z \, dz$$
where γ is a path joining 0 and $a + ib$. Equate the answer obtained with the parametric evaluation along the straight line from 0 to $a + ib$, and deduce that
$$\int_0^1 e^{ax} \cos(bx) dx = \frac{a(e^a \cos b - 1) + be^a \sin b}{a^2 + b^2}.$$

Exercise 3.15. Applying the Fundamental Theorem of Contour Integration to $\exp z$ and integrating round a circular path, show that for all $r > 0$,
$$\int_0^{2\pi} e^{r \cos \theta} \cos(r \sin \theta + \theta) d\theta = 0.$$

Exercise 3.16. Show that $1/z$ has no primitive in the punctured plane $\mathbb{C} \setminus \{0\}$.

3.4 The Cauchy Integral Theorem

We will now show one of the main results in complex analysis, called the Cauchy Integral Theorem.

Theorem 3.4. (The Cauchy Integral Theorem) *Let*

(1) *D be a domain in $\mathbb{C}$,*
(2) *$f : D \to \mathbb{C}$ be holomorphic in D, and*
(3) *$\gamma_0, \gamma_1 : [0,1] \to D$ be two closed, piecewise smooth, D-homotopic paths.*

Then $\displaystyle\int_{\gamma_0} f(z) \, dz = \int_{\gamma_1} f(z) \, dz.$

Before we go further, let us try to understand the statement. Notice, first of all, that the two paths in D are *closed*. Secondly, what do we mean by saying that the two closed paths are "D-homotopic"? Intuitively, this means the following. Look at Figure 3.12, where we have depicted the two paths in the domain. Imagine you have placed a rubber band along γ_0. For γ_0 to be D-homotopic to γ_1, we should be able to deform this rubber band so as to get γ_1, with the condition that each intermediate position of the rubber band lies in D. Clearly this is not possible sometimes, for example if the domain has holes. See for example the picture on the right of Figure 3.12, we expect the two paths in the domain D taken as punctured complex plane $D = \mathbb{C} \setminus \{0\}$ to be not D-homotopic.

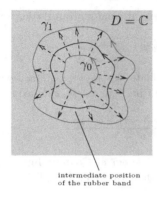
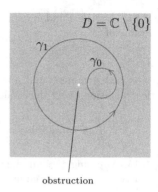

intermediate position
of the rubber band   obstruction

Fig. 3.12 γ_0, γ_1 in $D := \mathbb{C}$ are $\mathbb{C}$-homotopic in the picture on the left, but in the picture on the right, the two paths γ_0, γ_1 in $D = \mathbb{C} \setminus \{0\}$ are not $\mathbb{C} \setminus \{0\}$-homotopic.

We have the following precise definition.

Definition 3.3. Let D be a domain in $\mathbb{C}$ and $\gamma_0, \gamma_1 : [0,1] \to D$ be closed paths. Then γ_0 is said to be *D-homotopic to γ_1* if there is a continuous function $H : [0,1] \times [0,1] \to D$ such that the following hold:

(H1) For all $t \in [0,1]$, $H(t,0) = \gamma_0(t)$.
(H2) For all $t \in [0,1]$, $H(t,1) = \gamma_1(t)$.
(H3) For all $s \in [0,1]$, $H(0,s) = H(1,s)$.

We can think of the H as a family of closed paths from $[0,1]$ to D, parameterized by "time", the s-variable. You may think of the closed path at time s as the position of the rubber band at time s, Initially, when $s = 0$, $H(\cdot,0)$ is the path γ_0, while finally, when the time $s = 1$, we end up with $H(\cdot,1)$, which is the path γ_1. So far, this is what (H1) and (H2) say. The requirement (H3) just says that at each point of time s, the intermediate path $\gamma_s := H(\cdot,s)$ is closed too. Continuity of H means that the rubber band never breaks, and the deformation takes place smoothly. The picture below illustrates this.

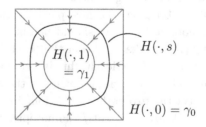

Example 3.8. Let $D = \mathbb{C}$, and $\gamma_0, \gamma_1 : [0,1] \to \mathbb{C}$ be the two circular paths given by $\gamma_0 = 4\exp(2\pi it)$, and $\gamma_1 = 2i + \exp(2\pi it)$, for $t \in [0,1]$. Then γ_0 is $\mathbb{C}$-homotopic to γ_1. Indeed, we can define H by just taking a "convex combination" of the points $\gamma_0(t)$ and $\gamma_1(t)$. See Figure 3.8.

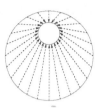

Set $H : [0,1] \times [0,1] \to \mathbb{C}$ by $H(t,s) = (1-s) \cdot \gamma_0(t) + s \cdot \gamma_1(t) = (4-3s) \cdot \exp(2\pi it) + s \cdot 2i$, $0 \le t, s \le 1$. Then H is clearly continuous, and moreover,

(H1) for all $t \in [0,1]$, $H(t,0) = (4-0) \cdot \exp(2\pi it) + 0 \cdot 2i = \gamma_0(t)$,
(H2) for all $t \in [0,1]$, $H(t,1) = (4-3) \cdot \exp(2\pi it) + 1 \cdot 2i = \gamma_1(t)$, and
(H3) for each $s \in [0,1]$, $H(0,s) = (4-3s) \cdot 1 + s \cdot 2i = H(1,s)$.

Hence (H1), (H2), (H3) are satisfied and so γ_0 is $\mathbb{C}$-homotopic to γ_1.

On the other hand, the same two paths are not $\mathbb{C} \setminus \{0\}$-homotopic. Why is that? If they were $\mathbb{C} \setminus \{0\}$-homotopic, then by the Cauchy Integral Theorem, the contour integral of the holomorphic function $1/z$ in $\mathbb{C} \setminus \{0\}$ along the two paths would be the same. But we have

$$\int_{\gamma_0} \frac{1}{z}\,dz = 2\pi i \ne 0 = \int_{\gamma_1} \frac{1}{z}\,dz,$$

where the last equality follows from the Fundamental Theorem of Contour Integration, because $1/z$ has the primitive $\mathrm{Log}\,z$ in $\mathbb{C} \setminus (-\infty, 0]$. ◊

Exercise 3.17. Let D be a domain in $\mathbb{C}$. Show that D-homotopy is an equivalence relation on the set of all closed paths in D. In particular, we can say "γ_0, γ_1 are D-homotopic" instead of saying that "γ_0 is D-homotopic to γ_1".

Proof. (of Theorem 3.4.) We will make the simplifying assumption that the homotopy H is twice continuously differentiable. This smoothness condition can be omitted (see for example [Conway (1978)]), but then the proof becomes technical. Moreover, the assumption of twice continuous differentiability is mild, and we will invoke this below when we will exchange the order of partial differentiation:

$$\frac{\partial^2 H}{\partial s \partial t} = \frac{\partial^2 H}{\partial t \partial s}.$$

Idea of the proof: Let $\gamma_s := H(\cdot, s)$ be the intermediate curve at time s. Define

$$I(s) := \int_{\gamma_s} f(z)dz, \quad s \in [0, 1].$$

(We will use differentiation under the integral sign with respect to s to show that

$$\frac{d}{ds}I(s) \equiv 0,$$

showing that $s \mapsto I(s)$ is constant, and in particular

$$\int_{\gamma_0} f(z)dz = I(0) \stackrel{?}{=} I(1) = \int_{\gamma_1} f(z)dz,$$

which is the desired conclusion.) We have

$$\frac{dI}{ds}(s) = \frac{d}{ds}\int_{\gamma_s} f(z)\,dz = \frac{d}{ds}\int_0^1 f(H(t, s))\frac{\partial H}{\partial t}(t, s)dt$$

$$= \int_0^1 \frac{\partial}{\partial s}\left(f(H(t, s))\frac{\partial H}{\partial t}(t, s) \right)dt$$

$$= \int_0^1 \left(f'(H(t, s))\frac{\partial H}{\partial s}(t, s)\frac{\partial H}{\partial t}(t, s) + f(H(t, s))\frac{\partial^2 H}{\partial s \partial t}(t, s) \right)dt$$

$$= \int_0^1 \left(f'(H(t, s))\frac{\partial H}{\partial t}(t, s)\frac{\partial H}{\partial s}(t, s) + f(H(t, s))\frac{\partial^2 H}{\partial t \partial s}(t, s) \right)dt$$

$$= \int_0^1 \frac{d}{dt}\left(f(H(t, s))\frac{\partial H}{\partial s}(t, s) \right)dt,$$

and so by the Fundamental Theorem of Integral Calculus,

$$\frac{dI}{ds}(s) = \int_0^1 \frac{d}{dt}\left(f(H(t, s))\frac{\partial H}{\partial s}(t, s) \right)dt$$

$$= f(H(1, s))\frac{\partial H}{\partial s}(1, s) - f(H(0, s))\frac{\partial H}{\partial s}(0, s)$$

$$= f(H(1, s))\lim_{\sigma \to s}\frac{H(1, \sigma) - H(1, s)}{\sigma - s} - f(H(1, s))\lim_{\sigma \to s}\frac{H(0, \sigma) - H(0, s)}{\sigma - s}$$

$$= f(H(1, s))\lim_{\sigma \to s}\frac{H(1, \sigma) - H(1, s)}{\sigma - s} - f(H(1, s))\lim_{\sigma \to s}\frac{H(1, \sigma) - H(1, s)}{\sigma - s}$$

$$= 0.$$

Hence the map $s \mapsto I(s) : [0, 1] \to \mathbb{C}$ is constant. In particular,

$$\int_{\gamma_1} f(z)\,dz = I(1) = I(0) = \int_{\gamma_2} f(z)\,dz.$$

This completes the proof. But where did we use the holomorphicity of f? We used this in order to get the equality in the third line above. Indeed, we claim that

$$\frac{\partial}{\partial s}(f(H(t,s))) = f'(H(t,s)) \cdot \frac{\partial H}{\partial s}(t,s).$$

Let $f = u + iv$, and $H = X + iY$, where u, v, X, Y are real valued. We suppress the arguments (t, s) below. Then we have

$$\frac{\partial}{\partial s}(f(H(t,s))) = \frac{\partial}{\partial s}(u(X,Y) + iv(X,Y))$$

$$= \frac{\partial u}{\partial x}(X,Y) \cdot \frac{\partial X}{\partial s} + \frac{\partial u}{\partial y}(X,Y) \cdot \frac{\partial Y}{\partial s}$$

$$+ i \left(\frac{\partial v}{\partial x}(X,Y) \cdot \frac{\partial X}{\partial s} + \frac{\partial v}{\partial y}(X,Y) \cdot \frac{\partial Y}{\partial s} \right)$$

$$= \frac{\partial u}{\partial x}(X,Y) \cdot \frac{\partial X}{\partial s} - \frac{\partial v}{\partial x}(X,Y) \cdot \frac{\partial Y}{\partial s}$$

$$+ i \left(\frac{\partial v}{\partial x}(X,Y) \cdot \frac{\partial X}{\partial s} + \frac{\partial u}{\partial x}(X,Y) \cdot \frac{\partial Y}{\partial s} \right)$$

$$= \left(\frac{\partial u}{\partial x}(X,Y) + i\frac{\partial v}{\partial x}(X,Y) \right) \cdot \left(\frac{\partial X}{\partial s} + i\frac{\partial Y}{\partial s} \right)$$

$$= f'(X + iY) \cdot \frac{\partial}{\partial s}(X + iY) = f'(H(t,s)) \cdot \frac{\partial H}{\partial s}(t,s).$$

The careful reader would have also noticed that we also assumed that f' is continuous when we differentiated under the integral sign. Again, the result holds without this assumption, but we will not do this here. The interested student can find the complete proof of the Cauchy Integral Theorem for example in [Conway (1978)]. □

Exercise 3.18. We have seen that if C is the circular path with center 0 and radius 1 traversed in the anticlockwise direction, then

$$\int_C \frac{1}{z}dz = 2\pi i.$$

Now consider the path S, comprising the four line segments which are the sides of the square with vertices $\pm 1 \pm i$, traversed anticlockwise. Draw a picture to convince yourself that S is $\mathbb{C} \setminus \{0\}$-homotopic to C. Evaluate parametrically the integral

$$\int_S \frac{1}{z}dz,$$

and confirm that the answer is indeed $2\pi i$.

Exercise 3.19. Let $a > 0$, $b > 0$, and $E : [0, 2\pi] \to \mathbb{C}$ be the elliptic path

$$E(t) = a \cos t + ib \sin t, \quad t \in [0, 2\pi].$$

By considering $\int_E \dfrac{1}{z} dz$, show that $\displaystyle\int_0^{2\pi} \dfrac{1}{a^2(\cos \theta)^2 + b^2(\sin \theta)^2} d\theta = \dfrac{2\pi}{ab}$.

3.4.1 *Special case: simply connected domains*

Consider a "degenerate" closed path, which is constant. That is, if D is a domain and $p \in D$, then consider $\gamma_p : [0, 1] \to D$ given by $\gamma_p(t) = p$, $t \in [0, 1]$. Then γ_p is closed, because $\gamma_p(0) = p = \gamma_p(1)$. For any continuous $f : D \to \mathbb{C}$, what is

$$\int_{\gamma_p} f(z) dz?$$

0, because $\gamma_p'(t) = 0$ for each $t \in [a, b]$ and

$$\int_{\gamma_p} f(z) dz = \int_0^1 f(\gamma_p(t)) \cdot \gamma_p'(t) dt = 0.$$

In light of this, we see that an important special case of the Cauchy Integral Theorem is obtained when we know that a closed path γ is D-homotopic to a *point* (that is, the constant path $\gamma_p(t) = p$ for all $t \in [0, 1]$). In this case we say that γ is D-*contractible*. Imagine placing a rubber band along γ, and then shrinking it to a point such that each intermediate position of the rubber band is in D. Indeed for a D-contractible path γ (which is D-homotopic to a point $p \in D$), and for a holomorphic function $f : D \to \mathbb{C}$, we have by the Cauchy Integral Theorem that

$$\int_{\gamma} f(z) dz = \int_{\gamma_p} f(z) dz = 0.$$

A domain in which *every* closed path is D-contractible is called *simply connected*.

For example, the domains $\mathbb{C}$, $\mathbb{D} := \{z \in \mathbb{C} : |z| < 1\}$, $\mathbb{C} \setminus (-\infty, 0]$, are all simply connected. For example if we take any p in the domains in first two cases, then the homotopy given by

$$H(t, s) := (1 - s)\gamma(t) + sp, \quad t, s \in [0, 1]$$

does the job. In the case when $D = \mathbb{C} \setminus (-\infty, 0]$, given any $\gamma : [0, 1] \to D$, we first choose any real $p > 0$ for example $p = 1$, and then use the same H as above. Note that none of the above domains have any "holes" in them. On the other hand, domains with holes are not simply connected. For example, the punctured complex plane $\mathbb{C} \setminus \{0\}$ is not simply connected. For example consider the circular path with center 0 and any positive radius r traversed once in the anticlockwise direction. We have seen that

$$\int_C \frac{1}{z} \, dz = 2\pi i.$$

But if C were $\mathbb{C} \setminus \{0\}$-contractible to a point in the punctured plane, we should have had

$$\int_C \frac{1}{z} \, dz = 0$$

by Cauchy's Integral Formula. So this means that C is not $\mathbb{C} \setminus \{0\}$-contractible to a point in $\mathbb{C} \setminus \{0\}$ and so $\mathbb{C} \setminus \{0\}$ is not simply connected. Similarly, one can show that the annulus

$$\{z \in \mathbb{C} : 1 < |z| < 2\}$$

is not simply connected. See Figure 3.13. The obstruction of the hole can be thought of as a nail or a pillar emanating from the plane, which prevents a rubber band encircling it from being shrunk to a point in the domain while always staying in the plane.

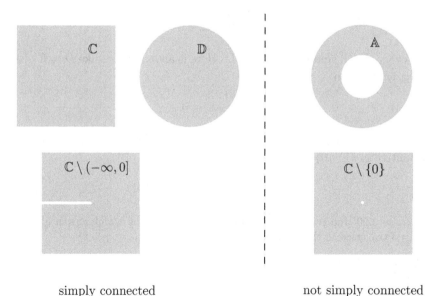

simply connected   not simply connected

Fig. 3.13   The domains $\mathbb{C}$, $\mathbb{D} := \{z \in \mathbb{C} : |z| < 1\}$ and $\mathbb{C} \setminus (-\infty, 0]$ are simply connected, while the annulus $\mathbb{A} := \{z \in \mathbb{C} : 1 < |z| < 2\}$ and the punctured plane $\mathbb{C} \setminus \{0\}$ aren't.

We have the following corollary of the Cauchy Integral Theorem.

Corollary 3.2. *Let*

(1) D *be a simply connected domain,*
(2) γ *be a closed piecewise smooth path in D and*
(3) $f : D \to \mathbb{C}$ *be holomorphic in D.*

Then $\displaystyle\int_\gamma f(z)\,dz = 0.$

This corollary itself is also sometimes called the Cauchy Integral Theorem.

Example 3.9. For any closed path γ, since exp is entire, and $\mathbb{C}$ is simply connected, we have

$$\int_\gamma \exp z\,dz = 0.$$

In fact for any entire function f, $\displaystyle\int_\gamma f(z)\,dz = 0$, for any closed path γ.   ◊

It can happen for a *non* simply connected domain D and a holomorphic function $f : D \to \mathbb{C}$ that

$$\int_\gamma f(z)dz = 0$$

for every closed smooth path γ in D. Here is an example: take $D = \mathbb{C}\backslash\{0\}$, $f := 1/z^2$. Then

$$\int_\gamma \frac{1}{z^2}dz = 0$$

for every closed path γ in the punctured plane because $1/z^2$ possesses a primitive in $\mathbb{C}\backslash\{0\}$:

$$\frac{d}{dz}\left(-\frac{1}{z}\right) = \frac{1}{z^2}, \quad z \in \mathbb{C}\backslash\{0\}.$$

Exercise 3.20. Integrate the following functions over the circular path given by $|z| = 3$ traversed in the anticlockwise direction:

(1) $\mathrm{Log}(z - 4i)$.

(2) $\dfrac{1}{z-1}$.

(3) Principal value of i^{z-3}.

Exercise 3.21. (Winding number of a curve.) Suppose that $\gamma : [0,1] \to \mathbb{C}$ is a smooth closed path that does not pass through 0. We define the *winding number of γ* (about 0) to be

$$w(\gamma) := \frac{1}{2\pi i} \int_\gamma \frac{1}{z} dz = \frac{1}{2\pi i} \int_0^1 \frac{\gamma'(t)}{\gamma(t)} dt.$$

(1) Using the observation that $\exp(2\pi i a) = 1$ if and only if $a \in \mathbb{Z}$, show that $w(\gamma) \in \mathbb{Z}$ by proceeding as follows. Define $\varphi : [0,1] \to \mathbb{C}$ by

$$\varphi(t) = \exp\left(\int_0^t \frac{\gamma'(s)}{\gamma(s)} ds\right), \quad t \in [0,1].$$

To show that $w(\gamma) \in \mathbb{Z}$, it suffices to show that $\varphi(1) = 1$. To this end, calculate $\varphi'(t)$, and use this expression to show that φ/γ is constant in $[0,1]$. Use this fact to conclude that $\varphi(1) = 1$.

(2) Calculate the winding number of $\Gamma_1 : [0,1] \to \mathbb{C}$ given by $\Gamma_1(t) = \exp(2\pi i t)$ ($t \in [0,1]$).

(3) Prove that if $\gamma_1, \gamma_2 : [0,1] \to \mathbb{C}$ are smooth closed paths not passing through 0, and $\gamma_1 \cdot \gamma_2$ is their pointwise product, then $w(\gamma_1 \cdot \gamma_2) = w(\gamma_1) + w(\gamma_2)$.

(4) Let $m \in \mathbb{N}$. Calculate the winding number of the curve $\Gamma_m : [0,1] \to \mathbb{C}$ given by $\Gamma_m(t) = \exp(2\pi i m t)$ ($t \in [0,1]$).

(5) Show that the winding number function $\gamma \mapsto w(\gamma)$ is "locally constant", by which we mean that if $\gamma_0 : [0,1] \to \mathbb{C} \setminus \{0\}$ is a smooth closed path, then there is a $\delta > 0$ such that for every smooth closed path $\gamma : [0,1] \to \mathbb{C} \setminus \{0\}$ such that $\|\gamma - \gamma_0\|_\infty := \max\{|\gamma(t) - \gamma_0(t)| : t \in [0,1]\} < \delta$, we have $w(\gamma) = w(\gamma_0)$. (In other words, if we equip the set of curves with the uniform topology, and equip $\mathbb{Z}$ with the discrete topology, then $\gamma \mapsto w(\gamma)$ is continuous.)

3.4.2 *What happens with nonholomorphic functions?*

We now highlight the fact that the Cauchy Integral Theorem may fail if one drops the assumption of holomorphicity of f. Let us see what happens when we consider our favourite nonholomorphic function, the complex conjugation map $z \mapsto \overline{z}$. We will show that rather than the integral around the closed loop γ being 0, the contour integral of $\overline{z}$ around γ yields the area enclosed by γ, which is of course very much dependent on γ, and two $\mathbb{C}$-homotopic paths can enclose widely different areas (just imagine two concentric circles with different radii). We will only give a plausibility argument by resorting to a specific picture, as shown in the picture below.

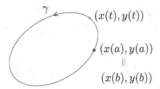

For the smooth path $\gamma : [a, b] \to \mathbb{C}$, we have

$$\int_\gamma \overline{z}\, dz$$

$$= \int_a^b (x(t) - iy(t))(x'(t) + iy'(t))dt$$

$$= \int_a^b x(t)x'(t) + y(t)y'(t) + i(x(t)y'(t) - y(t)x'(t))dt$$

$$= \frac{(x(b))^2 - (x(a))^2 + (y(b))^2 - (y(a))^2}{2} + i \int_a^b (x(t)y'(t) - y(t)x'(t))dt$$

$$= 0 + i \int_a^b (x(t)y'(t) - y(t)x'(t))dt.$$

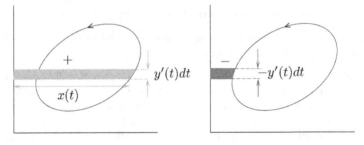

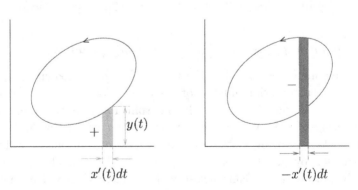

Fig. 3.14 $\displaystyle\int_a^b x(t)y'(t)dt$ and $\displaystyle\int_a^b x'(t)y(t)dt$.

Look at Figure 3.14. From the two pictures on the top, we see that

$$\int_a^b x(t)y'(t)dt = (\text{Area enclosed by } \gamma).$$

From the two pictures on the bottom, we see that

$$\int_a^b x'(t)y(t)dt = -(\text{Area enclosed by } \gamma).$$

Thus $\displaystyle\int_\gamma \overline{z}\,dz = i\int_a^b (x(t)y'(t) - y(t)x'(t))dt = 2i \cdot (\text{Area enclosed by } \gamma).$

So in contrast to the Cauchy Integral Theorem for holomorphic functions, we see that for this nonholomorphic function, the integral along a closed contour is not zero, but yields the area of the contour!

Exercise 3.22. Suppose a coin of radius r rolls around a fixed bigger coin of radius R. Then the path traced by a point on the rim of the rolling coin is called an *epicycloid*, and it is a closed curve if $R = nr$, for some $n \in \mathbb{N}$. See Figure 3.15.

Fig. 3.15   The epicycloid with $n = 6$.

(1) With the center of the fixed coin at the origin, show that the epicycloid can be represented parametrically as $z(t) = r((n + 1)\exp(it) - \exp(i(n + 1)t))$, $t \in [0, 2\pi]$.

(2) By evaluating the integral of $\overline{z}$ along the epicycloid, show that the area enclosed by the epicycloid is equal to $\pi r^2(n + 1)(n + 2)$.

In the rest of this chapter we will learn about several consequences of the Cauchy Integral Theorem. In particular, we will learn

(1) that in simply connected domains every holomorphic function possesses a primitive;

(2) that holomorphic functions are infinitely many times differentiable;

(3) that bounded entire functions are constants (Liouville's Theorem) (and also use this to prove the Fundamental Theorem of Algebra);

(4) a result called Morera's theorem, which is a sort of a converse to the Cauchy Integral Theorem.

3.5 Existence of a primitive

We will show that on a simply connected domain, every holomorphic function is the derivative of some holomorphic function.

Theorem 3.5. *If*

(1) *D is a simply connected domain and*
(2) *$f : D \to \mathbb{C}$ is holomorphic,*

then there is a holomorphic function $F : D \to \mathbb{C}$ such that for all $z \in D$,
$F'(z) = f(z)$.

Proof. Fix a point $p \in D$. Define $F : D \to \mathbb{C}$ by

$$F(z) = \int_{\gamma_z} f(\zeta)\,d\zeta, \quad z \in D,$$

where γ is *any* smooth path in D joining p to z. Why is this F well-defined, that is, why does the $F(z)$ not depend on *which* path we take joining p to z? If $\widetilde{\gamma}$ is another smooth path in D that joins p to z, then $\gamma - \widetilde{\gamma}$ is a closed smooth path in the simply connected domain D. See Figure 3.16.

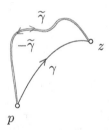

Fig. 3.16 $\gamma - \widetilde{\gamma}$ forms a closed path.

Cauchy Integral Theorem gives

$$0 = \int_{\gamma - \widetilde{\gamma}} f(z)\,dz = \int_{\gamma_z} f(\zeta)\,d\zeta - \int_{\widetilde{\gamma}} f(\zeta)\,d\zeta,$$

so that $\displaystyle\int_{\gamma_z} f(\zeta)\,d\zeta = \int_{\widetilde{\gamma}} f(\zeta)\,d\zeta$, and F is well-defined.

Next we show the holomorphicity of F and that $F' = f$ in D. Since f is holomorphic in D, it is also continuous there, and so given a $z \in D$ and an $\epsilon > 0$, there is a $\delta > 0$ such that whenever $|w - z| < \delta$, we have $|f(w) - f(z)| < \epsilon$. Thus if we take a w such that $0 < |w - z| < \delta$, then

$$\frac{F(w) - F(z)}{w - z} = \frac{1}{w - z}\left(\int_{\gamma_w} f(\zeta)\,d\zeta - \int_{\gamma_z} f(\zeta)\,d\zeta \right).$$

If γ_{zw} is a straight line path joining z to w, then the concatenation of γ_z with the concatenation of γ_{zw} with $-\gamma_w$ is a closed path, and so by the Cauchy Integral Theorem, we obtain

$$0 = \int_{\gamma_z + \gamma_{zw} - \gamma_w} f(\zeta)\,d\zeta = \int_{\gamma_z} f(\zeta)\,d\zeta + \int_{\gamma_{zw}} f(\zeta)\,d\zeta - \int_{\gamma_w} f(\zeta)\,d\zeta.$$

The Fundamental Theorem of Contour Integration gives

$$\int_{\gamma_{zw}} 1\,d\zeta = \int_{\gamma_{zw}} \zeta'\,d\zeta = w - z,$$

and so

$$\frac{F(w) - F(z)}{w - z} - f(z) = \frac{1}{w - z}\int_{\gamma_{zw}} f(\zeta)\,d\zeta - \frac{1}{w - z}\int_{\gamma_{zw}} f(z)\,d\zeta$$

$$= \frac{1}{w - z}\int_{\gamma_{zw}} (f(\zeta) - f(z))\,d\zeta,$$

and so

$$\left| \frac{F(w) - F(z)}{w - z} - f(z) \right| = \left| \frac{1}{w - z}\int_{\gamma_{zw}} (f(\zeta) - f(z))\,d\zeta \right|$$

$$= \frac{1}{|w - z|}\left| \int_{\gamma_{zw}} (f(\zeta) - f(z))\,d\zeta \right|$$

$$\leq \frac{1}{|w - z|}\left(\max_{\zeta \in \gamma_{zw}} |f(\zeta) - f(z)| \right) \cdot (\text{length of } \gamma_{zw})$$

$$\leq \frac{1}{|w - z|}\, \epsilon |w - z| = \epsilon.$$

Thus $F'(z) = f(z)$, and F is holomorphic. $\square$

Remark 3.2. A primitive for a holomorphic function f in a simply connected domain is unique up to a constant. Indeed, if $F, \widetilde{F}$ are both primitives for F, then $F' = f = \widetilde{F}'$ in D, and so

$$\frac{d}{dz}(F - \widetilde{F}) = F' - \widetilde{F}' = f - f = 0 \text{ in } D.$$

By Exercise 2.13, it follows that there is a constant C such that $F - \widetilde{F} = C$ in D. So $F = \widetilde{F} + C$ in D.

Example 3.10. $\exp(-z^2)$ is entire. So there exists an F, which is also entire, such that for all $z \in \mathbb{C}$, $F'(z) = \exp(-z^2)$. (But one cannot express F in terms of elementary functions. One primitive is given by

$$\widetilde{F}(z) = \int_{\gamma_z} e^{-\zeta^2} d\zeta$$

for $z \in \mathbb{C}$, where γ_z is the straight line path joining 0 to z. Then in particular, for real x,

$$\widetilde{F}(x) = \int_0^x e^{-\xi^2} d\xi,$$

and it turns out that this (and so any other primitive too) can't be expressed in terms of elementary functions.) $\Diamond$

Exercise 3.23. Suppose that D is a domain. If f is holomorphic in D, and there is no F holomorphic in D such that $F' = f$ in D, then we know that D cannot be simply connected. Give a concrete example of such a D and f.

Corollary 3.3. *If*

(1) *D is a simply connected domain,*
(2) *$f : D \to \mathbb{C}$ is holomorphic,*
(3) *$\gamma : [a, b] \to D$ and $\widetilde{\gamma} : [c, d] \to D$ are two smooth paths such that they have the same start and end points, that is, $\gamma(a) = \widetilde{\gamma}(c)$ and $\gamma(b) = \widetilde{\gamma}(d)$,*

then $\displaystyle\int_\gamma f(z)\,dz = \int_{\widetilde{\gamma}} f(z)\,dz.$

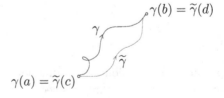

$$\gamma(b) = \widetilde{\gamma}(d)$$
$$\gamma(a) = \widetilde{\gamma}(c)$$

Proof. f has a primitive F and so

$$\int_\gamma f(z)\,dz = F(\gamma(1)) - F(\gamma(0)) = F(\widetilde{\gamma}(1)) - F(\widetilde{\gamma}(0)) = \int_{\widetilde{\gamma}} f(z). \qquad \square$$

3.6 The Cauchy Integral Formula

We will now learn about a result, called the Cauchy Integral Formula, which says, roughly speaking that if we have a closed path γ without self-intersections, and f is a function which is holomorphic inside γ, then the value of f at any point inside γ is determined by the values of the function on γ! This illustrates the "rigidity" of holomorphic functions.

Later on, in the next chapter, we will study a more general Cauchy Integral Formula, which will allow us to even express all the derivatives of f at any point inside γ in terms of the values of the function on γ. So we can consider the basic result in this section as the "$n = 0$ case" of the more general result to follow.

Theorem 3.6. (The Cauchy Integral Formula for circular paths)
Let

(1) *D be a domain,*
(2) *$f : D \to \mathbb{C}$ be holomorphic in D,*
(3) *$r > 0$, $z_0 \in D$ and the disc $\Delta := \{z \in \mathbb{C} : |z - z_0| \leq r\} \subset D$.*

Then

$$f(w) = \frac{1}{2\pi i} \int_{C_r} \frac{f(z)}{z - w} dz, \quad |w - z_0| < r,$$

where C_r is the circular path $C_r(t) = z_0 + r \exp(it)$, $t \in [0, 2\pi]$, with center z_0 and radius $r > 0$ traversed in the anticlockwise direction.

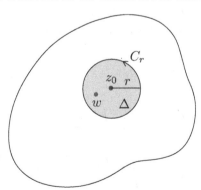

In order to prove this result, we will first prove the following technical fact, which will also prove to be useful later on.

Proposition 3.5. *Let*

(1) *D be a domain, $z_0 \in D$,*
(2) *$f : D \to \mathbb{C}$ be holomorphic in $D \setminus \{z_0\}$, and continuous on D,*
(3) *$r > 0$ and the disc $\Delta := \{z \in \mathbb{C} : |z - z_0| \leq r\}$ be contained in D.*

Then

$$f(z_0) = \frac{1}{2\pi i} \int_{C_r} \frac{f(z)}{z - z_0}\, dz,$$

where C_r is the circular path $C_r(t) = z_0 + r \exp(it)$, $t \in [0, 2\pi]$, with center z_0 and radius $r > 0$ traversed in the anticlockwise direction.

Proof. Let $\epsilon > 0$. Then there is a $\delta > 0$ (which we can arrange to be smaller than r) such that whenever $0 < |z - z_0| \leq \delta$, $|f(z) - f(z_0)| < \epsilon$. Consider the circular path C_δ, with center z_0 and radius δ traversed in the anticlockwise direction. But C_δ and C_r are easily seen to be $D \setminus \{z_0\}$-homotopic.

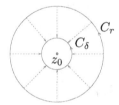

Indeed the homotopy H can be obtained by just taking the convex combination of the points on C_r and C_δ: $H(\cdot, s) := (1-s)C_r(\cdot) + sC_\delta(\cdot)$, $s \in [0, 1]$. Thus by the Cauchy Integral Theorem, we have

$$\int_{C_r} \frac{f(z)}{z - z_0}\, dz = \int_{C_\delta} \frac{f(z)}{z - z_0}\, dz.$$

Hence,

$$\left| \frac{1}{2\pi i} \int_{C_r} \frac{f(z)}{z - z_0}\, dz - f(z_0) \right| = \left| \frac{1}{2\pi i} \int_{C_\delta} \frac{f(z)}{z - z_0}\, dz - f(z_0) \frac{1}{2\pi i} \int_{C_\delta} \frac{1}{z - z_0}\, dz \right|$$

$$= \left| \frac{1}{2\pi i} \int_{C_\delta} \frac{f(z) - f(z_0)}{z - z_0}\, dz \right|$$

$$\leq \left(\max_{z \in C_\delta} \frac{|f(z) - f(z_0)|}{2\pi |z - z_0|} \right) \cdot 2\pi\delta$$

$$< \frac{\epsilon}{2\pi\delta} \cdot 2\pi\delta = \epsilon.$$

Since $\epsilon > 0$ was arbitrary, the claim follows. $\square$

The following is an immediate corollary.

Corollary 3.4. *Let*

(1) *D be a domain,*
(2) *$f : D \to \mathbb{C}$ be holomorphic in D,*
(3) *$r > 0$, $z_0 \in D$ and the disc $\Delta := \{z \in \mathbb{C} : |z - z_0| \leq r\} \subset D$.*

Then

$$f(z_0) = \frac{1}{2\pi i} \int_{C_r} \frac{f(z)}{z - z_0} dz,$$

where C_r is the circular path $C_r(t) = z_0 + r \exp(it)$, $t \in [0, 2\pi]$, with center z_0 and radius $r > 0$ traversed in the anticlockwise direction.

We now prove the basic version of the Cauchy Integral Formula, namely Theorem 3.6. Note that as opposed to the previous Corollary 3.4, now the w can be *any* point inside the circle with center z_0 and radius r, and not necessarily the center z_0 as in the corollary above.

Proof. (of Theorem 3.6.) Let w be such that $|w - z_0| < r$. Choose a $\delta > 0$ small enough so that the circular path C_δ with center w and radius δ is contained in the interior of C_r. But now C_r and C_δ are $D \setminus \{w\}$-homotopic, and this can be seen in the same manner as in Example 3.8.

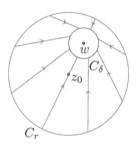

Since

$$\frac{f(\cdot)}{\cdot - w}$$

is holomorphic in $D \setminus \{w\}$, it follows from the Cauchy Integral Theorem that the second equality holds below:

$$f(w) = \frac{1}{2\pi i} \int_{C_\delta} \frac{f(z)}{z - w} dz = \frac{1}{2\pi i} \int_{C_r} \frac{f(z)}{z - w} dz.$$

This completes the proof. $\square$

Exercise 3.24. Let $0 < a < 1$, and let γ be the unit circle with center 0 traversed anticlockwise. Show that

$$\int_\gamma \frac{i}{(z-a)(az-1)}dz = \int_0^{2\pi} \frac{1}{1+a^2-2a\cos t}dt.$$

Use Cauchy's Integral Formula to deduce that $\displaystyle\int_0^{2\pi} \frac{1}{1+a^2-2a\cos t}dt = \frac{2\pi}{1-a^2}.$

Exercise 3.25. Fill in the blanks.

(1) $\displaystyle\int_\gamma \frac{\exp z}{z-1}dz =$ _____ , where γ is the circle $|z| = 2$ traversed in the anticlockwise direction.

(2) $\displaystyle\int_\gamma \frac{z^2+1}{z^2-1}dz =$ _____ , where γ is the circle $|z-1| = 1$ traversed in the anticlockwise direction.

(3) $\displaystyle\int_\gamma \frac{z^2+1}{z^2-1}dz =$ _____ , where γ is the circle $|z-i| = 1$ traversed in the anticlockwise direction.

(4) $\displaystyle\int_\gamma \frac{z^2+1}{z^2-1}dz =$ _____ , where γ is the circle $|z+1| = 1$ traversed in the anticlockwise direction.

(5) $\displaystyle\int_\gamma \frac{z^2+1}{z^2-1}dz =$ _____ , where γ is the circle $|z| = 3$ traversed in the anticlockwise direction.

Exercise 3.26. Does $z \mapsto \dfrac{1}{z(1-z^2)}$ have a primitive in $\{z \in \mathbb{C} : 0 < |z| < 1\}$?

Corollary 3.5. (Cauchy's Integral Formula for general paths)
Let

(1) *D be a domain,*
(2) *$f : D \to \mathbb{C}$ be holomorphic in D,*
(3) *$z_0 \in D$, and*
(4) *γ be a closed path in D which is $D \setminus \{z_0\}$-homotopic to a circular path C centered at z_0, such that C and its interior is contained in D.*

Then we have $f(z_0) = \dfrac{1}{2\pi i} \displaystyle\int_\gamma \frac{f(z)}{z-z_0}dz.$

Proof. By the Cauchy Integral Formula for circular paths, it follows that

$$f(z_0) = \frac{1}{2\pi i} \int_C \frac{f(z)}{z-z_0}dz.$$

But since γ is $D \setminus \{z_0\}$-homotopic to C, by the Cauchy Integral Theorem,

$$\frac{1}{2\pi i} \int_C \frac{f(z)}{z - z_0} dz = \frac{1}{2\pi i} \int_\gamma \frac{f(z)}{z - z_0} dz.$$

This completes the proof. □

This result highlights the "rigidity" associated with holomorphic functions mentioned earlier. By this we mean that their highly structured nature (everywhere locally infinitesimally a rotation followed by a magnification) enables one to pin down their precise behaviour from very limited information. That is, even if we know the effect of a holomorphic function in a small portion of the plane (for example the values along a closed path), its values can be inferred at other far away points in a unique manner. The picture below illustrates this in the case of the Cauchy Integral Formula, where knowing the values of f on the curve γ enables one to determine the values at all points in the shaded region!

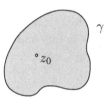

Exercise 3.27. Let F be defined by $F(z) = \dfrac{\exp(iz)}{z^2 + 1}$ and let $R > 1$.

(1) Let σ be the closed semicircular path formed by the segment S of the real axis from $-R$ to R, followed by the circular arc T of radius R in the upper half plane from R to $-R$. Show that

$$\int_\sigma F(z) dz = \frac{\pi}{e}.$$

(2) Prove that $|\exp(iz)| \leq 1$ for z in the upper half plane, and conclude that for large enough $|z|$, $|F(z)| \leq 2/|z|^2$.

(3) Show that $\lim\limits_{R \to \infty} \int_T F(z) \, dz = 0$, and so $\lim\limits_{R \to \infty} \int_S F(z) \, dz = \dfrac{\pi}{e}$.

(4) Conclude, by parameterizing the integral over S in terms of x, that

$$\int_{-\infty}^{\infty} \frac{\cos x}{1 + x^2} dx := \lim_{R \to \infty} \int_{-R}^{R} \frac{\cos x}{1 + x^2} dx = \frac{\pi}{e}.$$

Exercise 3.28. Evaluate $\displaystyle\int_0^{2\pi} e^{\cos \theta} \cos(\sin \theta) \, d\theta$. *Hint:* Consider $\exp(\exp(i\theta))$.

3.7   Holomorphic functions are infinitely differentiable

In this section we prove the fundamental property of holomorphic functions in a domain, namely that they are infinitely many times complex differentiable.

Let us contrast this with the situation in Real Analysis. We have already seen in Example 0.1 that the derivative may fail to be differentiable at isolated points. There are even more extreme examples of this phenomenon, and there exist functions $f : \mathbb{R} \to \mathbb{R}$ which are differentiable everywhere, but f' is differentiable nowhere! We refer the interested reader to §3.8, [Gelbaum and Olmsted (1964)], where one can find an example of a function $g : \mathbb{R} \to \mathbb{R}$ that is continuous everywhere, but differentiable nowhere; the integral

$$f(x) = \int_0^x g(\xi)d\xi, \quad x \in \mathbb{R},$$

of this g, then gives our sought for $f : \mathbb{R} \to \mathbb{R}$ that is differentiable everywhere, but whose derivative (g!) is differentiable nowhere.

Corollary 3.6. *Let*

(1) *D be a domain, and*
(2) *$f : D \to \mathbb{C}$ be holomorphic in D.*

Then f' is holomorphic in D.

Note that the above gives the following chain of implications:

$$\boxed{f \in \mathrm{Hol}(D)} \Rightarrow \boxed{f' \in \mathrm{Hol}(D)} \Rightarrow \boxed{f'' \in \mathrm{Hol}(D)} \Rightarrow \cdots$$

So whenever f is holomorphic in a domain D (that is, $f \in \mathrm{Hol}(D)$), it is infinitely many times complex differentiable.

Here is a plan of how we will show this. From the Cauchy Integral Formula, we know that

$$f(z) = \frac{1}{2\pi i} \int_{C_r} \frac{f(\zeta)}{\zeta - z} \, d\zeta,$$

where C_r is a circle centered at z with radius r. If we were to formally differentiate under the integral sign, we would get an expression for the derivative of f:

$$f'(z) = \frac{1}{2\pi i} \int_{C_r} \frac{f(\zeta)}{(\zeta - z)^2} \, d\zeta,$$

Having shown this formula, we will show that

$$\lim_{w \to z} \frac{f'(w) - f'(z)}{z - w}$$

exists by using the above expression for the derivative at z and w.

Proof. Let $z_0 \in D$. Let g be defined by

$$g(z) = \begin{cases} \dfrac{f(z) - f(z_0)}{z - z_0} & \text{if } z \neq z_0, \\ f'(z_0) & \text{if } z = z_0. \end{cases}$$

Clearly g is holomorphic in $D \setminus \{z_0\}$ and continuous in D. We will now apply the technical fact we had shown in Proposition 3.5 to g. Choose an $r > 0$ small enough so that the disc $\{z \in \mathbb{C} : |z - z_0| \leq r\}$ is contained in D, and let C_r denote the circular path with center z_0 and radius r traversed anticlockwise. Then

$$\begin{aligned} f'(z_0) = g(z_0) &= \frac{1}{2\pi i} \int_{C_r} \frac{g(z)}{z - z_0} \, dz \\ &= \frac{1}{2\pi i} \int_{C_r} \frac{f(z) - f(z_0)}{(z - z_0)^2} \, dz \qquad (3.4) \\ &= \frac{1}{2\pi i} \int_{C_r} \frac{f(z)}{(z - z_0)^2} \, dz - \frac{f(z_0)}{2\pi i} \int_{C_r} \frac{1}{(z - z_0)^2} \, dz \\ &= \frac{1}{2\pi i} \int_{C_r} \frac{f(z)}{(z - z_0)^2} \, dz - 0. \qquad (3.5) \end{aligned}$$

Thus for w inside C_r, but with $w \neq z_0$, we have that

$$f'(w) = \frac{1}{2\pi i} \int_{C_\delta} \frac{f(z)}{(z - w)^2} \, dz = \frac{1}{2\pi i} \int_{C_r} \frac{f(z)}{(z - w)^2} \, dz,$$

where C_δ is a small circular path with center w and radius δ that lies inside C_r. The second equality above follows from the Cauchy Integral Theorem, because

$$\frac{f(\cdot)}{(\cdot - w)^2}$$

is holomorphic in $D \setminus \{w\}$ and the paths C_r, C_δ are $D \setminus \{w\}$-homotopic.

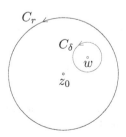

So we have for $w \neq z_0$ inside C_r that

$$\frac{f'(w) - f'(z_0)}{w - z_0} = \frac{1}{w - z_0} \left(\frac{1}{2\pi i} \int_{C_r} \frac{f(z)}{(z-w)^2} dz - \frac{1}{2\pi i} \int_{C_r} \frac{f(z)}{(z-z_0)^2} dz \right)$$

$$= \frac{1}{2\pi i} \int_{C_r} \frac{f(z)(2z - z_0 - w)}{(z-w)^2(z-z_0)^2} dz.$$

What does this look like when $w \approx z_0$? The numerator looks like

$$f(z)(2z - z_0 - z_0) = 2f(z)(z - z_0),$$

while the denominator looks like $(z - z_0)^2(z - z_0)^2$. So we guess that

$$\lim_{w \to z_0} \frac{f'(w) - f'(z_0)}{w - z_0} = \frac{2}{2\pi i} \int_{C_r} \frac{f(z)}{(z - z_0)^3} dz,$$

and we prove this claim below. So let us calculate

$$\frac{f'(w) - f'(z_0)}{w - z_0} - \frac{2}{2\pi i} \int_{C_r} \frac{f(z)}{(z - z_0)^3} dz = (w - z_0) \frac{1}{2\pi i} \int_{C_r} \frac{(3z - z_0 - 2w)f(z)}{(z-w)^2(z-z_0)^3} dz$$

Of course if we manage to show that the integral is bounded by some constant for all w close to z_0, then we see that since this is being multiplied by $w - z_0$, as $w \to z_0$, we can make the overall expression as small as we please. To this end, let us consider a disc with center z_0 and having radius smaller than r, say $r/2$, and we will confine w (which is anyway supposed to be near z_0) to lie within this disc. So from now on, w will lie in the (compact set)

$$\left\{ w \in \mathbb{C} : |w - z_0| \leq \frac{r}{2} \right\}.$$

Consider the continuous map φ,

$$(z, w) \overset{\varphi}{\mapsto} \left| \frac{(3z - z_0 - 2w)f(z)}{(z-w)^2(z-z_0)^3} \right|$$

$$: C_r \times \left\{ w \in \mathbb{C} : |w - z_0| \leq \frac{r}{2} \right\} (=: K \subset \mathbb{C}^2 = \mathbb{R}^4) \to \mathbb{R}.$$

But K is a compact set in $\mathbb{R}^4$ because it is closed and bounded. Thus the continuous function $\varphi : K \to \mathbb{R}$ has some maximum value $M \geq 0$ on the compact set K. Hence,

$$\left| \frac{1}{2\pi i} \int_{C_r} \frac{(3z - z_0 - 2w)f(z)}{(z-w)^2(z-z_0)^3} dz \right| \leq \frac{1}{2\pi} M (\text{length of } C_r) = \frac{1}{2\pi} M 2\pi r = Mr,$$

and so

$$\left| \frac{f'(w) - f'(z_0)}{w - z_0} - \frac{2}{2\pi i} \int_{C_r} \frac{f(z)}{(z - z_0)^3} dz \right| \leq |w - z_0| Mr \overset{w \to z_0}{\longrightarrow} 0.$$

This shows that f' is differentiable at z_0. As the choice of z_0 was arbitrary, f' is holomorphic in D. $\qquad \square$

Exercise 3.29. Suppose f is holomorphic in a domain D. Is it clear that if $n \in \mathbb{N}$, then $f^{(n)}$ has a continuous complex derivative?

3.8 Liouville's Theorem; Fundamental Theorem of Algebra

Here is one more instance of the rigidity associated with holomorphicity.

Theorem 3.7. (Liouville's Theorem) *Every bounded entire function is constant.*

Let us again contrast this with the situation in Real Analysis. There we know that for example $x \mapsto \sin x$ is differentiable everywhere on $\mathbb{R}$, and it is bounded too: $|\sin x| \leq 1$ for all $x \in \mathbb{R}$. But sin is not a constant function.

On the other hand, in light of the above Liouville's Theorem, the entire function $z \mapsto \sin z$, being nonconstant, must necessarily be unbounded on $\mathbb{C}$! We had checked this by brute force using the definition earlier on page 23, and proved that $|\sin(iy)| \to \infty$ as $y \to \pm\infty$.

Proof. Let $M \geq 0$ be such that for all $z \in \mathbb{C}$, $|f(z)| \leq M$. Suppose that $w \in \mathbb{C}$, and let γ be the circular path with center w and radius R, where R is any positive number. Then (from the proof of Corollary 3.6, see in particular (3.5))

$$f'(w) = \frac{1}{2\pi i} \int_\gamma \frac{f(z)}{(z-w)^2} dz,$$

and so

$$|f'(w)| = \left| \frac{1}{2\pi i} \int_\gamma \frac{f(z)}{(z-w)^2} dz \right| \leq \frac{1}{2\pi} \cdot \frac{M}{R^2} \cdot 2\pi R = \frac{M}{R}.$$

But since $R > 0$ was arbitrary, it follows that $f'(w) = 0$. So $f'(w) = 0$ for all $w \in \mathbb{C}$, and hence f is constant. We had seen this in Exercise 2.13, but here is another way to see this. If $z \in \mathbb{C}$, by considering the straight line path γ_z joining 0 to z, we have

$$f(z) - f(0) = \int_{\gamma_z} f'(\zeta) d\zeta = 0.$$

This completes the proof. $\qquad \square$

This result can be used to give a short proof of the Fundamental Theorem of Algebra[2].

Corollary 3.7. (Fundamental Theorem of Algebra) *Every polynomial of degree ≥ 1 has a root in $\mathbb{C}$.*

---

[2]Despite its name, there is no purely algebraic proof of the theorem, since any proof must use the completeness of the reals, which is not an algebraic concept. Additionally, it is not really "fundamental" for modern algebra; its name was given at a time when the study of algebra was mainly concerned with the solutions of polynomial equations with real or complex coefficients.

For a polynomial $p : \mathbb{C} \to \mathbb{C}$, given by $p(z) = c_0 + c_1 z + \cdots + c_d z^d$, $z \in \mathbb{C}$, where $c_0, c_1, \cdots, c_d \in \mathbb{C}$, $c_d \neq 0$, the number d is called the *degree of p*. A number $z_0 \in \mathbb{C}$ such that $p(z_0) = 0$ is called a *zero/root of p*.

Proof. Suppose $p(z) = c_0 + c_1 z + \cdots + c_d z^d$ is a polynomial with $d \geq 1$, and such that it has no root in $\mathbb{C}$. That is, for all $z \in \mathbb{C}$, $p(z) \neq 0$. But since p is entire and nonzero, its reciprocal, namely the function f given by $f(z) = 1/p(z)$ ($z \in \mathbb{C}$), is entire; see Exercise 2.6. In Exercise 1.24, we had shown an estimate on the growth of a polynomial:

there exist $M, R > 0$ such that $|p(z)| \geq M|z|^d$ whenever $|z| > R$.

In the compact set $\{z \in \mathbb{C} : |z| \leq R\}$, the continuous function $z \mapsto |p(z)|$ has a positive (because p is never 0) minimum m. Thus

$$|f(z)| \leq \min\left\{\frac{1}{MR^d}, \frac{1}{m}\right\}, \quad z \in \mathbb{C}.$$

By Liouville's Theorem, f must be constant, and so p must be a constant, a contradiction to the fact that $d \geq 1$. ∎

For example, the polynomial p given by $p(z) = z^{1976} - 3z^{28} + \sqrt{399}$ is guaranteed to have a root somewhere in the complex plane.

Exercise 3.30. Let f be an entire function such that f is bounded away from 0, that is, there is a $\delta > 0$ such that for all $z \in \mathbb{C}$, $|f(z)| \geq \delta$. Show that f is a constant.

Exercise 3.31. Show that an entire function whose range of values avoids a disc $\{w \in \mathbb{C} : |w - w_0| < r\}$ must be a constant.

Exercise 3.32. Assume that f is an entire function that is periodic in both the real and in the imaginary direction, that is, there exist T_1, T_2 in $\mathbb{R}$ such that $f(z) = f(z + T_1) = f(z + iT_2)$ for all $z \in \mathbb{C}$. Prove that f is constant.

Exercise 3.33. A classical theme in the theory of entire functions is to try to characterize the entire function f based on the way $|f|$ grows for large $|z|$. Here is one instance of this.

(1) Show that if f is entire and $|f(z)| \leq |\exp z|$, for all $z \in \mathbb{C}$, then in fact f is equal to $c \cdot \exp z$ for some complex constant c with $|c| \leq 1$. (Thus if a nonconstant entire function "grows" no faster than the exponential function, it *is* an exponential function.)

(2) One may be tempted to argue that this can't be right on the grounds that "polynomials grow more slowly than the exponential function", but surely $p \neq \exp z$. Find the flaw in this reasoning by showing that if p is a polynomial satisfying $|p(z)| \leq |\exp z|$ for all $z \in \mathbb{C}$, then $p \equiv 0$.

Hint: Look at $z = x < 0$.

Exercise 3.34. Let $f : \mathbb{C} \to \mathbb{C}$ be an entire function. Suppose that a_1, a_2 are complex numbers such that $a_1 \neq a_2$, and such that a_1, a_2 are contained in the interior of the circular path C with radius $R > 0$ and center 0, traversed once in the counterclockwise direction.

(1) Prove that $\left| \displaystyle\int_C \dfrac{f(z)}{(z - a_1)(z - a_2)} dz \right| \leq \dfrac{2\pi R}{(R - |a_1|)(R - |a_2|)} \max_{z \in C} |f(z)|.$

(2) Find $\alpha, \beta \in \mathbb{C}$ such that for all $z \in \mathbb{C}$, $\dfrac{1}{(z - a_1)(z - a_2)} = \dfrac{\alpha}{z - a_1} + \dfrac{\beta}{z - a_2}.$

(3) Express $\displaystyle\int_C \dfrac{f(z)}{(z - a_1)(z - a_2)} dz$ in terms of $\displaystyle\int_C \dfrac{f(z)}{z - a_1} dz$ and $\displaystyle\int_C \dfrac{f(z)}{z - a_2} dz.$

Use the Cauchy Integral Formula to simplify these latter expressions.

(4) Deduce Liouville's Theorem.

3.9 Morera's Theorem: converse to Cauchy's Integral Theorem

Recall the Cauchy Integral Theorem, where we have learnt that if

(1) D is a domain,
(2) $f : D \to \mathbb{C}$ is holomorphic,
(3) Δ is any disc such that $\Delta \subset D$,
(4) γ is a closed piecewise smooth path in Δ,

then $\displaystyle\int_\gamma f(z) dz = 0.$

Now we will see that the following converse to the above result holds.

Theorem 3.8. (Morera's Theorem) *If*

(1) *D is a domain,*
(2) *$f : D \to \mathbb{C}$ is a continuous function such that*
(3) *for every closed rectangular path γ in every disc contained in D,*
$$\int_\gamma f(z) dz = 0,$$

then f is holomorphic in D.

In other words if the contour integral is zero for some special paths, then we are allowed to conclude that f is holomorphic!

Proof. Let $z_0 \in D$, and let Δ be a disc with center z_0 such that $\Delta \subset D$, and $\gamma_{z_0,z}$ is the path joining z_0 to z by first moving horizontally and then moving vertically.

Define $F : \Delta \to \mathbb{C}$ by $F(z) = \displaystyle\int_{\gamma_{z_0,z}} f(\zeta)\, d\zeta, \quad z \in \Delta.$

We will show that F is holomorphic in Δ, and its derivative is f. This shows that f is holomorphic in Δ! Why? f (being the derivative of a holomorphic function) is itself holomorphic in Δ.

Let $z \in \Delta$. Suppose $\epsilon > 0$. Since f is continuous, there exists $\delta > 0$ such that whenever $|w - z| < \delta$ and $w \in \Delta$, it follows that $|f(w) - f(z)| < \epsilon$. We have

$$F(w) - F(z) = \int_{\gamma_{z_0,w}} f(\zeta)\, d\zeta - \int_{\gamma_{z_0,z}} f(\zeta)\, d\zeta.$$

Using the fact that the integral of f on closed rectangular paths is zero,

$$F(w) - F(z) = \int_{\gamma_{z,w}} f(\zeta)\, d\zeta,$$

where $\gamma_{z,w}$ is the path joining z to w by again first moving horizontally and then moving vertically. See the picture below which shows one particular case, and we have (suppressing the integrand):

$$F(w) - F(z) = \int_{\gamma_{z_0,w}} f(\zeta)\, d\zeta - \int_{\gamma_{z_0,z}} f(\zeta)\, d\zeta = \int_A + \int_B + \int_C - \left(\int_A + \int_D \right)$$

$$= \underbrace{\int_B + \int_C + \int_{-\gamma_{z,w}} + \int_{-D}}_{=0} + \int_{\gamma_{z,w}} f(\zeta)\, d\zeta.$$

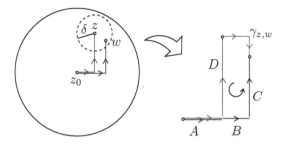

Thus for $0 < |w - z| < \delta$,

$$\frac{F(w) - F(z)}{w - z} - f(z) = \frac{1}{w - z} \int_{\gamma_{z,w}} f(\zeta)\, d\zeta - f(z)\frac{1}{w - z} \int_{\gamma_{z,w}} 1\, d\zeta$$

$$= \frac{1}{w - z} \int_{\gamma_{z,w}} (f(\zeta) - f(z))\, d\zeta,$$

where we have used the Fundamental Theorem of Contour Integration for the holomorphic function 1 to obtain

$$\int_{\gamma_{z,w}} 1\, d\zeta = w - z.$$

Consequently, from the above, and using the fact that the length of $\gamma_{z,w}$ is

$$|\mathrm{Re}(w - z)| + |\mathrm{Im}(w - z)| \leq 2|w - z|,$$

we have

$$\left| \frac{F(w) - F(z)}{w - z} - f(z) \right| = \left| \frac{1}{w - z} \int_{\gamma_{z,w}} (f(\zeta) - f(z))\, d\zeta \right|$$

$$\leq \frac{\epsilon}{|w - z|}(|\mathrm{Re}(w - z)| + |\mathrm{Im}(w - z)|) < 2\epsilon.$$

This completes the proof. $\qquad\square$

3.10   Notes

The proof of Theorem 3.4 follows closely the exposition in [Beck, Marchesi, Pixton, Sabalka (2008)]. Exercises 3.3, 3.4, 3.12, 3.20, 3.27 are taken from [Beck, Marchesi, Pixton, Sabalka (2008)]. Exercises 3.10, 3.14, 3.18, 3.19, 3.22, 3.24 are taken from [Needham (1997)]. Exercise 3.21 is taken from [Rudin (1987)]. Exercise 3.26, 3.33, 3.34 is taken from [Flanigan (1972)]. Exercise 3.28 is taken from [Howie (2003)].

Chapter 4

Taylor and Laurent series

In this chapter we will first learn about the fundamental result which says that a holomorphic function f has a power series expansion around any point in the domain D where it lives. See the picture on the left below.

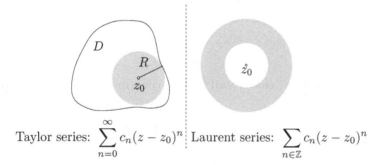

Taylor series: $\displaystyle\sum_{n=0}^{\infty} c_n (z - z_0)^n$ Laurent series: $\displaystyle\sum_{n \in \mathbb{Z}} c_n (z - z_0)^n$

That is, for each $z_0 \in D$, there exists an $R > 0$ such that

$$f(z) = \sum_{n=0}^{\infty} c_n (z - z_0)^n, \quad |z - z_0| < R.$$

Vice versa, every power series

$$\sum_{n=0}^{\infty} c_n (z - z_0)^n$$

that converges for at least two points converges in $|z - z_0| < R$ for some R and is holomorphic there. En route we will also prove further fundamental results on holomorphic functions:

(1) The (general) Cauchy Integral Formula and the Cauchy inequality.
(2) The classification of zeros and the Identity Theorem.
(3) The Maximum Modulus Theorem.

In the second part of the chapter, we will learn about Laurent series, which are like power series, except that negative integer powers of the terms $z - z_0$ also occur in the expansion. This will be useful to study functions that are holomorphic in annuli (and in particular punctured discs). See the picture on the right above. They are also useful to classify "singularities", and to evaluate some real integrals, as we will see at the end of this chapter.

4.1 Series

Just like with real series, given a sequence $(a_n)_{n \in \mathbb{N}}$ of complex numbers, one can form a new sequence $(s_n)_{n \in \mathbb{N}}$ of its *partial sums*:

$$s_1 := a_1,$$
$$s_2 := a_1 + a_2,$$
$$s_3 := a_1 + a_2 + a_3,$$
$$\vdots$$

Definition 4.1.

(1) The series $\displaystyle\sum_{n=1}^{\infty} a_n$ *converges* if $(s_n)_{n \in \mathbb{N}}$ converges, and $\displaystyle\sum_{n=1}^{\infty} a_n := \lim_{n \to \infty} s_n$.

(2) The series $\displaystyle\sum_{n=1}^{\infty} a_n$ *diverges* if $(s_n)_{n \in \mathbb{N}}$ diverges.

(3) $\displaystyle\sum_{n=1}^{\infty} a_n$ *converges absolutely* if the real series $\displaystyle\sum_{n=1}^{\infty} |a_n|$ converges.

From the result in Exercise 1.25, which says that a complex sequence converges if and only if the sequences of its real and imaginary parts converge,

$$\sum_{n=1}^{\infty} a_n \text{ converges} \Leftrightarrow \text{ the real series } \sum_{n=1}^{\infty} \mathrm{Re}(a_n) \text{ and } \sum_{n=1}^{\infty} \mathrm{Im}(a_n) \text{ converge.}$$

Thus the results from real analysis lend themselves for use in testing the convergence of complex series. For example, it is easy to prove the following two facts, which we leave as exercises.

Exercise 4.1. If $\displaystyle\sum_{n=1}^{\infty} a_n$ converges, then $\displaystyle\lim_{n \to \infty} a_n = 0$.

Exercise 4.2. If $\displaystyle\sum_{n=1}^{\infty} a_n$ converges absolutely, then $\displaystyle\sum_{n=1}^{\infty} a_n$ converges.

Exercise 4.3. Show that if $|z| < 1$, then $\displaystyle\sum_{n=0}^{\infty} z^n$ converges and that $\displaystyle\sum_{n=0}^{\infty} z^n = \frac{1}{1-z}$.

Exercise 4.4. Show that if $|z| < 1$, then $\displaystyle\sum_{n=1}^{\infty} nz^{n-1} = \frac{1}{(1-z)^2}$.

Exercise 4.5. Show that the series $1^{-s} + 2^{-s} + 3^{-s} + \cdots$ converges for all $s \in \mathbb{C}$ satisfying $\mathrm{Re}(s) > 1$. Thus

$$s \mapsto \zeta(s) := \sum_{n=1}^{\infty} \frac{1}{n^s}$$

is a well-defined map in the half-plane given by $\mathrm{Re}(s) > 1$, and is called the *Riemann zeta function*. The link of the Riemann zeta function with the number theoretic world of primes is brought out by the *Euler Product Formula*, which says that if $p_1 := 2 < p_2 := 3 < p_3 := 5 < \cdots$ is the infinite list of primes in increasing order, then

$$\zeta(s) = \lim_{K \to \infty} \prod_{k=1}^{K} \frac{1}{1 - p_k^{-s}}, \quad \mathrm{Re}(s) > 1.$$

Bernhard Riemann (1826-1866) showed that the function ζ can be extended holomorphically to $\mathbb{C} \setminus \{1\}$. It can be shown that the function ζ has zeros at $-2, -4, -6, \ldots$, called "trivial zeros", but it also has other zeros. All the non-trivial zeros Riemann computed turned out to lie on the line $\mathrm{Re}(s) = 1/2$. This led him to formulate the following conjecture, which is a famous unsolved problem in Mathematics.

Conjecture 4.1. (Riemann Hypothesis) *All non-trivial zeros of the Riemann zeta function lie on the line* $\mathrm{Re}(s) = 1/2$.

4.2 Power series

4.2.1 *Power series and their region of convergence*

Let $(c_n)_{n \in \mathbb{N}}$ be a complex sequence (thought of as a sequence of "coefficients"). An expression of the type

$$\sum_{n=0}^{\infty} c_n z^n$$

is called a *power series* in the complex variable z. Thus we imagine putting in specific values of z in the above series. Then for some $z \in \mathbb{C}$, the power series will converge, while for other values of z it may diverge.

Example 4.1. All polynomial expressions are power series, with only finitely many nonzero coefficients. Polynomials converge for *all* $z \in \mathbb{C}$.

The power series

$$\sum_{n=0}^{\infty} z^n$$

converges whenever $|z| < 1$. It diverges if $|z| \geq 1$ since $\neg\left(\lim_{n \to \infty} z^n = 0\right)$. $\Diamond$

A fundamental question is:

For what values of $z \in \mathbb{C}$ does the power series $\displaystyle\sum_{n=0}^{\infty} c_n z^n$ converge?

The following result gives the answer to this question.

Theorem 4.1. *For* $\displaystyle\sum_{n=0}^{\infty} c_n z^n$, *exactly one of the following hold:*

(1) *Either it is absolutely convergent for all* $z \in \mathbb{C}$.

(2) *Or there is a unique nonnegative real number* R *such that*

   (a) $\displaystyle\sum_{n=0}^{\infty} c_n z^n$ *is absolutely convergent for all* $z \in \mathbb{C}$ *with* $|z| < R$, *and*

   (b) $\displaystyle\sum_{n=0}^{\infty} c_n z^n$ *is divergent for all* $z \in \mathbb{C}$ *with* $|z| > R$.

(The unique $R > 0$ in the above theorem is called the *radius of convergence* of the power series, and if the power series converges for all $z \in \mathbb{C}$, we say that the power series has an infinite radius of convergence, and write "$R = \infty$".)

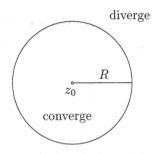

Fig. 4.1   Convergence region of a power series in $\mathbb{C}$.

What happens on the circle $|z| = R$? Complex power series may diverge at every point on the boundary (given by $|z| = R$), or diverge on some points of the boundary and converge at other points of the boundary, or converge at all the points on the boundary. There is no general result answering what happens at each point on the circle, and one just has to look at the specific power series at hand to find out the behaviour.

Proof. (of Theorem 4.1.) Let

$$S := \left\{ y \in [0, \infty) : \exists z \in \mathbb{C} \text{ such that } y = |z| \text{ and } \sum_{n=0}^{\infty} c_n z^n \text{ converges} \right\}.$$

Clearly $0 \in S$. Only two cases are possible:

$\underline{1°}$ S is not bounded above. In this case we will show that the radius of convergence is infinite. Given $z \in \mathbb{C}$, there exists a $y \in S$ such that $|z| < y$. But as $y \in S$, there exists a $z_0 \in \mathbb{C}$ such that $y = |z_0|$ and

$$\sum_{n=0}^{\infty} c_n z_0^n$$

converges. It follows that its terms tend to 0 as $n \to \infty$, and in particular, they are bounded: $|c_n z_0^n| \leq M$. Then with $r := |z|/|z_0|$ (< 1), we have

$$|c_n z^n| = |c_n z_0^n| \left(\frac{|z|}{|z_0|} \right)^n \leq M r^n \quad (n \in \mathbb{N}).$$

But $\sum_{n=0}^{\infty} M r^n$ converges $(r < 1!)$, and so by the Comparison Test,

$$\sum_{n=0}^{\infty} c_n z^n$$

is absolutely convergent. Since z was arbitrary, the claim follows.

$\underline{2°}$ S is bounded above. In this case, we will show that the radius of convergence is $\sup S$, that is,

(a) if $|z| < \sup S$, then $\sum_{n=0}^{\infty} c_n z^n$ converges absolutely, and

(b) if $|z| > \sup S$, then $\sum_{n=0}^{\infty} c_n z^n$ diverges.

If $z \in \mathbb{C}$ and $|z| < \sup S$, then by the definition of supremum, it follows that there exists a $y \in S$ such that $|z| < y$. Then we repeat the proof in 1° as follows. Since $y \in S$, there exists a $z_0 \in \mathbb{C}$ such that $y = |z_0|$ and

$$\sum_{n=0}^{\infty} c_n z_0^n$$

converges. It follows that its terms tend to 0 as $n \to \infty$, and in particular, they are bounded: $|c_n z_0^n| \leq M$. Then with $r := |z|/|z_0|$ (< 1), we have

$$|c_n z^n| = |c_n z_0^n| \left(\frac{|z|}{|z_0|} \right)^n \leq M r^n \quad (n \in \mathbb{N}).$$

But $\sum_{n=0}^{\infty} M r^n$ converges ($r < 1$!), and so $\sum_{n=0}^{\infty} c_n z^n$ is absolutely convergent.

Finally, if $z \in \mathbb{C}$ and $|z| > \sup S$, then setting $y := |z|$, we see that $y \notin S$, and by the definition of S,

$$\sum_{n=0}^{\infty} c_n z^n$$

diverges (otherwise $y \in S$).

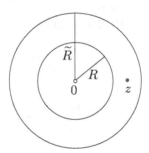

The uniqueness of R can be seen as follows. If $R, \tilde{R}$ have the property described in the theorem and $R < \tilde{R}$, then

$$R < r := \frac{R + \tilde{R}}{2} < \tilde{R}.$$

As $r < \tilde{R}$, $\sum_{n=1}^{\infty} c_n r^n$ converges. As $R < r$, $\sum_{n=0}^{\infty} c_n r^n$ diverges, a contradiction. □

The calculation of the radius of convergence is facilitated in some cases by the following result.

Theorem 4.2. *Consider the power series*

$$\sum_{n=0}^{\infty} c_n z^n.$$

If $L := \lim_{n \to \infty} \left| \dfrac{c_{n+1}}{c_n} \right|$ *exists, then*

(1) *the radius of convergence is* $1/L$ *if* $L \neq 0$.
(2) *the radius of convergence is infinite if* $L = 0$.

Proof. Let $L \neq 0$. We have that for all nonzero z such that $|z| < 1/L$ that there exists a $q < 1$ and an N large enough such that

$$\frac{|c_{n+1} z^{n+1}|}{|c_n z^n|} = \left| \frac{c_{n+1}}{c_n} \right| |z| \leq q < 1$$

for all $n > N$. (This is because

$$\left| \frac{c_{n+1}}{c_n} z \right| \overset{n \to \infty}{\longrightarrow} L|z| < 1.$$

So we may take $q = (L|z| + 1)/2 < 1$.) Thus by the Ratio Test, the power series converges absolutely for such z.

If $L = 0$, then for any nonzero $z \in \mathbb{C}$, we can guarantee that there exists a $q < 1$ such that

$$\frac{|c_{n+1} z^{n+1}|}{|c_n z^n|} = \left| \frac{c_{n+1}}{c_n} \right| |z| \leq q < 1$$

for all $n > N$. (This is because

$$\left| \frac{c_{n+1}}{c_n} z \right| \overset{n \to \infty}{\longrightarrow} 0|z| = 0 < 1.$$

So we may take $q = 1/2 < 1$.) Thus again by the Ratio Test, the power series converges absolutely for such z.

On the other hand, if $L \neq 0$ and $|z| > 1/L$, then there exists an N large enough such that

$$\frac{|c_{n+1} z^{n+1}|}{|c_n z^n|} = \left| \frac{c_{n+1}}{c_n} \right| |z| > 1 \text{ for all } n > N,$$

as $\left| \dfrac{c_{n+1}}{c_n} z \right| \overset{n \to \infty}{\longrightarrow} L|z| > 1$. By the Ratio Test, the power series diverges. $\qquad \square$

Example 4.2. Consider the power series $\displaystyle\sum_{n=1}^{\infty} \frac{z^n}{n^2}$. We have

$$\lim_{n \to \infty} \frac{\dfrac{1}{(n+1)^2}}{\dfrac{1}{n^2}} = 1,$$

and so the power series converges for $|z| < 1$ and diverges for $|z| > 1$. Note that if $|z| = 1$, then

$$\left| \frac{z^n}{n^2} \right| = \frac{1}{n^2},$$

and since $\displaystyle\sum_{n=1}^{\infty} \frac{1}{n^2}$ converges, it follows that

$$\sum_{n=1}^{\infty} \frac{z^n}{n^2}$$

converges absolutely. Thus at every point of the circle $|z| = 1$, the power series converges. We see that this situation is in contrast to the case of the geometric series

$$\sum_{n=0}^{\infty} z^n,$$

where we had convergence at no point of the circle $|z| = 1$.   ◊

Exercise 4.6. Consider the power series $\displaystyle\sum_{n=0}^{\infty} c_n x^n$. If $L := \lim_{n \to \infty} \sqrt[n]{|c_n|}$ exists, then

(1) the radius of convergence is $1/L$ if $L \neq 0$.

(2) the radius of convergence is infinite if $L = 0$.

Exercise 4.7. Show that $\displaystyle\sum_{n=1}^{\infty} n^n z^n$ converges only when $z = 0$.

Exercise 4.8. Show that $\displaystyle\sum_{n=1}^{\infty} \frac{z^n}{n^n}$ converges for all $z \in \mathbb{C}$.

Exercise 4.9. Find the radius of convergence of the following complex power series:

$$\sum_{n=1}^{\infty} \frac{(-1)^n}{n} z^n, \quad \sum_{n=0}^{\infty} n^{2012} z^n, \quad \sum_{n=0}^{\infty} \frac{1}{n!} z^n.$$

4.2.2 *Power series are holomorphic*

We have seen that polynomials are power series with an infinite radius of convergence, that is, they converge in the whole of $\mathbb{C}$. They are of course also holomorphic there. This is not a coincidence. Now we will see, more generally, that a power series

$$f(z) := \sum_{n=0}^{\infty} c_n z^n$$

that converges for $|z| < R$ is actually holomorphic there, and for $|z| < R$, there holds that

$$f'(z) = \frac{d}{dz}(c_0 + c_1 z + c_2 z^2 + \cdots) = c_1 + 2c_2 z + 3c_3 z^2 + \cdots = \sum_{n=1}^{\infty} c_n n z^{n-1},$$

(as expected, if one imagines differentiating the series termwise, as we do in the case of finite sums, that is, polynomials).

Theorem 4.3. *Let $R > 0$ and $f(z) := \sum_{n=0}^{\infty} c_n z^n$ converge for $|z| < R$. Then*

$$f'(z) = \sum_{n=1}^{\infty} n c_n z^{n-1} \quad for\ |z| < R.$$

Proof. **Step 1.** First we show that the power series

$$g(z) := \sum_{n=1}^{\infty} n c_n z^{n-1} = c_1 + 2c_2 z + \cdots + n c_n z^{n-1} + \ldots$$

is absolutely convergent for $|z| < R$. Fix z and let r satisfy $|z| < r < R$. By hypothesis

$$\sum_{n=0}^{\infty} c_n r^n$$

converges, and so there is some positive number M such that $|c_n r^n| < M$ for all n. Let $\rho := |z|/r$. Then $0 \le \rho < 1$, and

$$|n c_n z^{n-1}| = |c_n r^n| \cdot \frac{1}{r} \cdot n \left|\frac{z}{r}\right|^{n-1} \le \frac{M n \rho^{n-1}}{r}.$$

$\sum_{n=1}^{\infty} n \rho^{n-1}$ converges (to $1/(1-\rho)^2$; Exercise 4.4). By the Comparison Test, $\sum_{n=1}^{\infty} n c_n z^{n-1}$ converges absolutely.

Step 2. Now we show that $f'(z_0) = g(z_0)$ for $|z_0| < R$, that is,

$$\lim_{z \to z_0} \left(\frac{f(z) - f(z_0)}{z - z_0} - g(z_0) \right) = 0.$$

As before, let r be such that $|z_0| < r < R$ and since $z \to z_0$, we may also restrict z so that $|z| < r$.

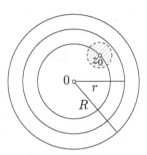

Let $\epsilon > 0$. As $\displaystyle\sum_{n=1}^{\infty} n c_n r^{n-1}$ converges absolutely, there is an N such that

$$\sum_{n=N}^{\infty} |n c_n r^{n-1}| < \frac{\epsilon}{4}.$$

Keep N fixed. We have $f(z) - f(z_0) = \displaystyle\sum_{n=1}^{\infty} c_n (z^n - z_0^n)$, and so for $z \neq z_0$,

$$\frac{f(z) - f(z_0)}{z - z_0} = \sum_{n=1}^{\infty} c_n \frac{z^n - z_0^n}{z - z_0} = \sum_{n=1}^{\infty} c_n \left(z^{n-1} + z^{n-2} z_0 + \cdots + z_0^{n-1} \right).$$

Thus

$$\frac{f(z) - f(z_0)}{z - z_0} - g(z_0) = \sum_{n=1}^{\infty} c_n \left(z^{n-1} + z^{n-2} z_0 + \cdots + z_0^{n-1} - n z_0^{n-1} \right).$$

We let S_1 be the sum of the first $N - 1$ terms of this series (that is, from $n = 1$ to $n = N-1$) and S_2 be the sum of the remaining terms. Then since $|z|, |z_0| < r$, it follows that

$$|S_2| \leq \sum_{n=N}^{\infty} |c_n| \Big(\underbrace{r^{n-1} + r^{n-1} + \cdots + r^{n-1}}_{n \text{ terms}} + n r^{n-1} \Big) = \sum_{n=N}^{\infty} 2n |c_n| r^{n-1} < \frac{\epsilon}{2}.$$

Also,

$$S_1 = \sum_{n=1}^{N} c_n \left(z^{n-1} + z^{n-2} z_0 + \cdots + z z_0^{n-2} + z_0^{n-1} - n z_0^{n-1} \right)$$

is a polynomial in z and by the algebra of limits,

$$\lim_{z \to z_0} S_1 = \sum_{n=1}^{N} c_n \left(z_0^{n-1} + z_0^{n-2} z_0 + \cdots + z_0 z_0^{n-2} + z_0^{n-1} - n z_0^{n-1} \right)$$

$$= \sum_{n=1}^{N} c_n \left(n z_0^{n-1} - n z_0^{n-1} \right) = 0.$$

So there is a $\delta > 0$ such that whenever $|z - z_0| < \delta$, we have $|S_1| < \epsilon/2$. Thus for $|z| < r$ and $0 < |z - z_0| < \delta$, we have

$$\left| \frac{f(z) - f(z_0)}{z - z_0} - g(z_0) \right| \leq |S_1| + |S_2| < \frac{\epsilon}{2} + \frac{\epsilon}{2} = \epsilon.$$

This means that $f'(z_0) = g(z_0)$, as claimed. □

Remark 4.1. If $(c_n)_{n \in \mathbb{N}}$ is a sequence of real numbers, then consider the *real* power series

$$\sum_{n=0}^{\infty} c_n x^n.$$

From Real Analysis, we know that such a power series converges in an interval of the form $(-R, R)$ and diverges in $\mathbb{R} \setminus [-R, R]$ for some $R \geq 0$. The two results in Theorems 4.1 and 4.3 imply that if we replace the real variable x by a complex variable z, then we can "extend/continue" the real power series to a holomorphic function in the disc given by $|z| < R$ in the complex plane. So we can view *real analytic* functions (namely functions of a real variable having a local power series expansion) as restrictions of holomorphic functions. This again highlights the interplay between the worlds of real analysis and complex analysis. (We have seen a previous instance of this interaction when we studied the Cauchy-Riemann equations.)

By a repeated application of the previous result, we have the following.

Corollary 4.1. *Let $R > 0$ and let $f(z) := \sum_{n=0}^{\infty} c_n z^n$ converge for $|z| < R$. Then for $k \geq 1$,*

$$f^{(k)}(z) = \sum_{n=k}^{\infty} n(n-1)(n-2) \cdots (n-k+1) c_n z^{n-k} \quad \text{for } |z| < R. \quad (4.1)$$

In particular, for $n \geq 0$, $c_n = \dfrac{1}{n!} f^{(n)}(0)$.

Proof. This is straightforward, and the last claim follows by setting $z = 0$ in (4.1):

$$f^{(k)}(0) = k(k-1)\cdots 1 c_k + z \sum_{n=k+1}^{\infty} n(n-1)\cdots(n-k+1)c_n z^{n-k-1} \Big|_{z=0} = k! c_k.$$

Also, $f(0) = c_0$. $\qquad\square$

There is nothing special about taking power series centered at 0. One can also consider

$$\sum_{n=0}^{\infty} c_n(z - z_0)^n,$$

where z_0 is a fixed complex number. The following results follow immediately from Theorems 4.1 and 4.3.

Corollary 4.2. *For* $\displaystyle\sum_{n=0}^{\infty} c_n(z - z_0)^n$, *exactly one of the following hold:*

(1) *Either it is absolutely convergent for all* $z \in \mathbb{C}$.

(2) *Or there is a unique nonnegative real number* R *such that*

    (a) $\displaystyle\sum_{n=0}^{\infty} c_n(z - z_0)^n$ *is absolutely convergent for* $|z - z_0| < R$, *and*

    (b) $\displaystyle\sum_{n=0}^{\infty} c_n(z - z_0)^n$ *is divergent for* $|z - z_0| > R$.

Corollary 4.3. *Let* $z_0 \in \mathbb{C}$, $R > 0$ *and* $f(z) := \displaystyle\sum_{n=0}^{\infty} c_n(z - z_0)^n$ *converge for* $|z - z_0| < R$. *Then*

$$f^{(k)}(z) = \sum_{n=k}^{\infty} n(n-1)\cdots(n-k+1)c_n(z-z_0)^{n-k} \text{ for } |z - z_0| < R, \ \ k \geq 1.$$

In particular, for $n \geq 0$, $c_n = \dfrac{1}{n!} f^{(n)}(z_0)$.

Remark 4.2. (Uniqueness of coefficients.) Suppose that

$$\sum_{n=0}^{\infty} c_n(z - z_0)^n \text{ and } \sum_{n=0}^{\infty} \tilde{c}_n(z - z_0)^n$$

are two power series which both converge to the same function f in an open disk with center z_0 and radius $R > 0$. Then from the above, for $n \geq 0$, we have

$$c_n = \frac{f^{(n)}(z_0)}{n!} = \tilde{c}_n.$$

Exercise 4.10. For $|z| < 1$, what is $1^2 + 2^2 z + 3^2 z^2 + 4^2 z^3 + \cdots$?

Exercise 4.11. True or false? All statements refer to power series $\sum_{n=0}^{\infty} c_n z^n$.

(1) The set of points z for which the power series converges equals either the singleton set $\{0\}$ or some open disc of finite positive radius or the entire complex plane, but no other type of set.

(2) If the power series converges for $z = 1$, then it converges for all z with $|z| < 1$.

(3) If the power series converges for $z = 1$, then it converges for all z with $|z| = 1$.

(4) If the power series converges for $z = 1$, then it converges for $z = -1$.

(5) Some power series converge at all points of an open disc with center 0 of some positive radius, and also at certain points on the boundary of the disc (that is, the circle bounding the disc), and at no other points.

(6) There are power series that converge on a set of points which is exactly equal to the closed disc given by $|z| \leq 1$.

(7) If the power series diverges at $z = i$, then it diverges at $z = 1 + i$ as well.

4.3 Taylor series

We have seen in the last section that complex power series

$$\sum_{n=0}^{\infty} c_n (z - z_0)^n$$

are holomorphic in their region of convergence $|z - z_0| < R$, where R is the radius of convergence. In this section, we will show that conversely, if f is holomorphic in the disc $|z - z_0| < R$, then

$$f(z) = \sum_{n=0}^{\infty} c_n (z - z_0)^n \text{ whenever } |z - z_0| < R,$$

where the coefficients can be determined from the f. Thus every holomorphic function f defined in a domain D possesses a power series expansion in a disc around any point $z_0 \in D$.

Theorem 4.4. *If f is holomorphic in $D(z_0, R) := \{z \in \mathbb{C} : |z - z_0| < R\}$, then $f(z) = c_0 + c_1(z - z_0) + c_2(z - z_0)^2 + c_3(z - z_0)^3 + \cdots$ for $z \in D(z_0, R)$, where for $n \geq 0$,*

$$c_n = \frac{1}{2\pi i} \int_C \frac{f(\zeta)}{(\zeta - z_0)^{n+1}} d\zeta,$$

and C is the circular path with center z_0 and radius r, where $0 < r < R$ traversed in the anticlockwise direction.

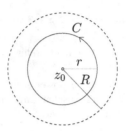

Proof. Let $z \in D(z_0, R)$. Initially, let r be such that $|z - z_0| < r < R$. Then by Cauchy's Integral Formula,

$$f(z) = \frac{1}{2\pi i} \int_C \frac{f(\zeta)}{\zeta - z} d\zeta = \frac{1}{2\pi i} \int_C \frac{f(\zeta)}{\zeta - z_0 + z_0 - z} d\zeta$$

$$= \frac{1}{2\pi i} \int_C \frac{f(\zeta)}{(\zeta - z_0)\left(1 - \dfrac{z - z_0}{\zeta - z_0}\right)} d\zeta$$

Set $w := \dfrac{z - z_0}{\zeta - z_0}$. Then $|w| = \dfrac{|z - z_0|}{r} < 1$. Thus

$$\frac{1}{1 - \dfrac{z - z_0}{\zeta - z_0}} = \frac{1}{1 - w} = 1 + w + w^2 + w^3 + \cdots + w^{n-1} + \frac{w^n}{1 - w}$$

$$= 1 + \frac{z - z_0}{\zeta - z_0} + \cdots + \frac{(z - z_0)^{n-1}}{(\zeta - z_0)^{n-1}} + \frac{(z - z_0)^n}{(\zeta - z_0)^{n-1}(\zeta - z)},$$

and so plugging this in the above, we obtain

$$f(z) = \frac{1}{2\pi i} \int_C f(\zeta) \left(\frac{1}{\zeta - z_0} + \cdots + \frac{(z - z_0)^{n-1}}{(\zeta - z_0)^n} + \frac{(z - z_0)^n}{(\zeta - z_0)^n(\zeta - z)} \right) d\zeta$$

$$= c_0 + c_1(z - z_0) + \cdots + c_{n-1}(z - z_0)^{n-1} + R_n(z),$$

where

$$R_n(z) := \frac{1}{2\pi i} \int_C \frac{f(\zeta)(z - z_0)^n}{(\zeta - z_0)^n(\zeta - z)} d\zeta.$$

So we would be done if we manage to show that $R_n(z)$ goes to 0 as $n \to \infty$. We note that $|f|$ is bounded on the circle, since it is a continuous real valued function on the compact set C, that is, there is an $M > 0$ such that for all $\zeta \in C$, $|f(\zeta)| < M$. (Here with a slight abuse, we think of the path C, and the set of points $C(t)$, $t \in [0, 2\pi]$, as being the same.) Also, for $\zeta \in C$,

$$\left| \frac{(z - z_0)^n}{(\zeta - z_0)^n} \right| = \left(\frac{|z - z_0|}{r} \right)^n \overset{n \to \infty}{\longrightarrow} 0.$$

But what about $1/|\zeta - z|$ for $\zeta \in C$? Is this bounded by something? The picture below shows that indeed this term is bounded by the reciprocal of the "distance between the circle C and z".

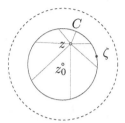

We have $|\zeta - z| = |\zeta - z_0 - (z - z_0)| \geq |\zeta - z_0| - |z - z_0| = r - |z - z_0|$. Thus

$$|R_n(z)| \leq \left(\frac{|z - z_0|}{r} \right)^n \frac{M}{r - |z - z_0|} \xrightarrow{n \to \infty} 0.$$

Thus the series $c_0 + c_1(z - z_0) + c_2(z - z_0)^2 + c_3(z - z_0)^3 + \cdots$ converges to $f(z)$. Note that we have only shown the expression

$$c_n = \frac{1}{2\pi i} \int_C \frac{f(\zeta)}{(\zeta - z_0)^{n+1}} d\zeta$$

where r is such that $|z - z_0| < r < R$. But by the Cauchy Integral Theorem, we see that this integral is independent of r, and any value of $r \in (0, R)$ can be chosen here:

(1) $\dfrac{f(\cdot)}{(\cdot - z_0)^{n+1}}$ is holomorphic in the punctured disc $D_*(z_0, R)$ given by $0 < |z - z_0| < R$;

(2) besides C, if $\widetilde{C}$ is another circular path with center z_0 and some other radius $\widetilde{r} \in (0, R)$ then $C, \widetilde{C}$ are $D_*(z_0, R)$-homotopic.

This completes the proof. $\qquad\qquad\qquad\qquad\qquad\qquad\qquad\qquad\qquad\square$

But we had learnt earlier in Theorem 4.3 that whenever we have a

$$f(z) = \sum_{n=0}^{\infty} c_n(z - z_0)^n \text{ for } |z - z_0| < R,$$

we know that

$$c_n = \frac{f^{(n)}(z_0)}{n!}$$

for $n \geq 0$. And in the above result, we found different expressions for the coefficients c_n, given in terms of integrals. But the coefficients of the power series expansion are unique in any disc, and so these have to be the same. With this observation in mind, we obtain the following result.

Corollary 4.4. (Taylor[1] Series) *If*

(1) *D be a domain,*
(2) *$f : D \to \mathbb{C}$ is holomorphic, and*
(3) *$z_0 \in D$,*

then

$$f(z) = f(z_0) + \frac{f'(z_0)}{1!}(z - z_0) + \frac{f''(z_0)}{2!}(z - z_0)^2 + \ldots, \quad |z - z_0| < R,$$

where R is the radius of the largest open disk with center z_0 contained in D. Also,

$$f^{(n)}(z_0) = \frac{n!}{2\pi i} \int_C \frac{f(z)}{(z - z_0)^{n+1}} dz, \tag{4.2}$$

where C is the circular path with center z_0 and radius r, where $0 < r < R$ traversed in the anticlockwise direction.

(4.2) is called the (general) *Cauchy Integral formula*. Earlier we had just seen the cases when $n = 0$ (Theorem 3.4) and $n = 1$ (Proof of Theorem 3.6). Also, we can obtain from here that

$$f^{(n)}(w) = \frac{n!}{2\pi i} \int_C \frac{f(z)}{(z - w)^{n+1}} dz,$$

for any point $w \in \Delta := \{z \in \mathbb{C} : |z - z_0| < R\}$. This follows from the Cauchy Integral Theorem, since we can first consider a small circle C_δ centered at w for which the above formula holds, and then note that the paths C, C_δ are $\Delta \setminus \{w\}$-homotopic, while the function $f(\cdot)/(\cdot - w)^{n+1}$ is holomorphic in $\Delta \setminus \{w\}$.

Proof. From Theorem 4.4, we have

$$f(z) = c_0 + c_1(z - z_0) + c_2(z - z_0)^2 + c_3(z - z_0)^3 + \ldots, \quad z \in D(z_0, R), \tag{4.3}$$

where $D(z_0, R) := \{z \in \mathbb{C} : |z - z_0| < R\}$ and R is the radius of the largest open disk with center z_0 contained in D. Also, for $n \geq 0$,

$$c_n = \frac{1}{2\pi i} \int_C \frac{f(z)}{(z - z_0)^{n+1}} dz,$$

---

[1]Named after Brook Taylor (1685-1731) who, among others, studied this expansion in the context of real analytic functions.

and C is the circular path with center z_0 and radius r, where $0 < r < R$ traversed in the anticlockwise direction. But from Corollary 4.3, the power series above is infinitely many times differentiable, and also, for $n \geq 0$,

$$\frac{1}{n!} f^{(n)}(z_0) = c_n.$$

Thus the result follows.　　　□

Summarizing, if f is holomorphic in $|z - z_0| < R$, then

$$f(z) = \sum_{n=0}^{\infty} \left(\frac{1}{2\pi i} \int_C \frac{f(\zeta)}{(\zeta - z_0)^{n+1}} d\zeta \right) \cdot (z - z_0)^n = \sum_{n=0}^{\infty} \frac{f^{(n)}(z_0)}{n!} (z - z_0)^n,$$

for $|z - z_0| < R$, where $C(t) = z_0 + r \exp(it)$, $t \in [0, 2\pi]$, and r is any number such that $0 < r < R$.

Example 4.3. The exponential function f, $z \mapsto f(z) := \exp z$, is entire. So we know that we can write

$$\exp z = \sum_{n=0}^{\infty} c_n z^n \text{ for } z \in \mathbb{C},$$

for some coefficients c_ns. What are these coefficients? They are given by

$$c_n = \frac{f^{(n)}(z_0)}{n!}, \quad n \geq 0.$$

Since $\dfrac{d}{dz} \exp z = \exp z$, it follows that $f^{(n)}(0) = 1$, and so

$$f(z) = \sum_{n=0}^{\infty} \frac{f^{(n)}(0)}{n!} (z - 0)^n = \sum_{n=0}^{\infty} \frac{1}{n!} z^n,$$

for all $z \in \mathbb{C}$.　　　◊

Example 4.4. The function f defined by $f(z) = \mathrm{Log}(z)$ is holomorphic in $\mathbb{C} \setminus (-\infty, 0]$. The largest open disc with center $z_0 = 1$ in this cut plane is $D = \{z \in \mathbb{C} : |z - 1| < 1\}$. Since

$$f^{(n)}(z_0) = \frac{(-1)^n (n - 1)!}{z_0^n} = (-1)^n (n - 1)!,$$

we have $\mathrm{Log}(1 + w) = w - \dfrac{w^2}{2} + \cdots + \dfrac{(-1)^n w^n}{n} + \cdots$ for $|w| < 1$.　　　◊

Exercise 4.12. Show that for $z \in \mathbb{C}$:

$$\sin z = z - \frac{z^3}{3!} + \frac{z^5}{5!} - + \cdots \text{ and } \cos z = 1 - \frac{z^2}{2!} + \frac{z^4}{4!} - + \cdots .$$

Exercise 4.13. Find the Taylor series of the polynomial $z^6 - z^4 + z^2 - 1$ with the point $z_0 = 1$ taken as the center.

Exercise 4.14. Find c_n's in the Taylor series $\displaystyle\sum_{n=0}^{\infty} c_n z^n$ around 0 of f given by

(1) $f(z) = \displaystyle\int_{\gamma_{0z}} \exp(\zeta^2)d\zeta$, $z \in \mathbb{C}$, where γ_{0z} is the straight line path from 0 to z.

    Hint: $f'(z) = \exp(z^2)$.

(2) $f(z) = \dfrac{z^2}{(z+1)^2}$, $z \in \mathbb{C}\backslash\{-1\}$. *Hint:* For $|z| < 1$, $\dfrac{1}{z+1} = 1 - z + z^2 - + \cdots$.

Here is a consequence of the general Cauchy Integral Formula.

Corollary 4.5. (Cauchy's inequality) *If*

(1) *f is holomorphic in $D(z_0, R) := \{z \in \mathbb{C} : |z - z_0| < R\}$ and*
(2) *$|f(z)| \leq M$ for all $z \in D(z_0, R)$,*

then for $n \geq 0$, $|f^{(n)}(z_0)| \leq \dfrac{n!M}{R^n}$.

Proof. Let C be the circle with center z_0 and radius $r < R$. Then

$$
\begin{aligned}
|f^{(n)}(z_0)| &= \left| \frac{n!}{2\pi i} \int_C \frac{f(z)}{(z - z_0)^{n+1}} dz \right| \\
&\leq \frac{n!}{2\pi} \max_{z \in C} \left| \frac{f(z)}{(z - z_0)^{n+1}} \right| \cdot 2\pi r = \frac{n!}{2\pi} \frac{M}{r^{n+1}} 2\pi r = \frac{n!M}{r^n}.
\end{aligned}
$$

The claim now follows by passing the limit $r \nearrow R$. $\qquad\qquad\square$

Exercise 4.15. Suppose that f is an entire function for which there is an $M > 0$ and an integer $n \geq 0$ such that for all $z \in \mathbb{C}$, $|f(z)| \leq M|z|^n$. Use Cauchy's inequality to prove that $f^{(n+1)}(z) = 0$ for all z and show that f is a polynomial of degree at most n. What happens when $n = 0$?

Exercise 4.16. Evaluate $\displaystyle\int_C \frac{\sin z}{z^{2013}} dz$, where C is the circular path with center 0 and radius 1 traversed in the anticlockwise direction.

4.4 Classification of zeros

Suppose $f : D \to \mathbb{C}$ is holomorphic in a domain D. We ask the question: What do the zeros of a nonzero f look like? (A point $z_0 \in D$ is a *zero* of f if $f(z_0) = 0$.) We will learn in this section that the answer to this question is, that the zeros are "isolated". Such a thing doesn't happen with continuous functions. Zeros of nonzero continuous functions, which aren't as rigid as holomorphic functions needn't be isolated.

Example 4.5.

(1) $\exp z$ has no zeros in $\mathbb{C}$. Indeed, $|\exp(z)| = e^{\mathrm{Re}(z)} > 0$ for all $z \in \mathbb{C}$.
(2) $\cos z - 3$ has infinitely many zeros in $\mathbb{C}$ at $2\pi n \pm i \log(3 + 2\sqrt{2})$, $n \in \mathbb{Z}$, all of which lie on a horizontal line, and they are isolated, with a distance of 2π between any two isolated adjacent zeros. We had seen this in Exercise 1.38.
(3) The polynomial p, $p(z) = (z + 1)^3 z^9 (z - 1)^9$, has zeros at $-1, 0, 1$. $\Diamond$

If p is a nonzero polynomial such that $p(z_0) = 0$, then by the Division Algorithm there exists a polynomial q (the quotient) such that

$$p(z) = (z - z_0)q(z)$$

(that is, the remainder is 0). Now we have two possible cases:

1° $q(z_0) \neq 0$. Then z_0 is an isolated zero of p.
2° $q(z_0) = 0$. Then we repeat the above procedure with p replaced by q.

Eventually, we obtain $p(z) = (z - z_0)^m q(z)$ for some $m \geq 1$ and $q(z_0) \neq 0$. Then we we call m the *multiplicity/order of z_0* (as a zero of p). We will now see in Theorem 4.1 below that the same sort of a thing holds for holomorphic functions f (replacing the polynomial p), except that we end up with another holomorphic function (g instead of the polynomial q). This is not completely surprising, since we know that power series are analogues of polynomials, and every holomorphic function has a local power series expansion.

But let us first give the following definitions.

Definition 4.2. Let D be a domain and $f : D \to \mathbb{C}$ be holomorphic in D. A point $z_0 \in D$ is called a *zero of f* if $f(z_0) = 0$.
If there is a smallest $m \in \mathbb{N}$ such that

(1) $f^{(m)}(z_0) \neq 0$ and
(2) $f(z_0) = \cdots = f^{(m-1)}(z_0) = 0$,

then z_0 is said to be *a zero of f of order m*. (We adopt the convention that $f^{(0)} := f$.)

We have the following result on the classification of zeros of a holomorphic function.

Proposition 4.1. (Classification of zeros) *Let*

(1) D *be a domain,*
(2) $f : D \to \mathbb{C}$ *be holomorphic in D and*
(3) $z_0 \in D$ *be a zero of f.*

Then there are exactly two possibilities:

$\underline{1^\circ}$ *There is a positive R such that $f(z) = 0$ for all z satisfying $|z - z_0| < R$.*
$\underline{2^\circ}$ *There exists an $m \in \mathbb{N}$ such that z_0 is a zero of f of order m, and there exists a holomorphic function $g : D \to \mathbb{C}$ such that $g(z_0) \neq 0$ and $f(z) = (z - z_0)^m g(z)$ for all $z \in D$.*

We note that if we are in case 2°, then since g is continuous and as $g(z_0) \neq 0$, g is not zero in a small disc Δ centered at z_0, and so f is nonzero in $\Delta \setminus \{0\}$ using the fact that for $z \in \Delta \setminus \{z_0\}$, $f(z) = (z - z_0)^m g(z) \neq 0$. Thus z_0 is the *only* zero of f in Δ, that is, z_0 is isolated.

Also, we will soon learn about something called the Identity Theorem, and from that result, it will follow that in fact in case 1°, we can conclude that $f \equiv 0$ in the whole of the domain D.

Proof. We have a power series expansion for f in a disc with some radius $R > 0$ and center z_0:
$$f(z) = c_0 + c_1(z - z_0) + c_2(z - z_0)^2 + \cdots \text{ for } |z - z_0| < R.$$
Since $f(z_0) = 0$, we know that $c_0 = 0$. Now there are exactly two possibilities:

$\underline{1^\circ}$ All the c_n are zero. Then $f(z) = 0$ whenever $|z - z_0| < R$.
$\underline{2^\circ}$ There is a smallest $m \geq 1$ such that $c_m \neq 0$. Then we have that $c_0 = c_1 = \cdots = c_{m-1} = 0$, and so using the fact that
$$c_n = \frac{f^{(n)}(z_0)}{n!}, \quad n \geq 0,$$
it follows that z_0 is a zero of order m. Moreover, from the power series expansion, we have
$$f(z) = c_m(z - z_0)^m + c_{m+1}(z - z_0)^{m+1} + \cdots = (z - z_0)^m \sum_{k=0}^{\infty} c_{m+k}(z - z_0)^k$$
$$\tag{4.4}$$
for $|z - z_0| < R$. Thus, if we define $g : D \to \mathbb{C}$ by
$$g(z) = \begin{cases} \dfrac{f(z)}{(z - z_0)^m} & \text{for } z \neq z_0, \\ \displaystyle\sum_{k=0}^{\infty} c_{m+k}(z - z_0)^k & \text{for } |z - z_0| < R, \end{cases}$$

From (4.4), the two definitions give the same value whenever both are applicable. So g is well-defined. Moreover, we have:

(1) g is holomorphic in D. For $z \neq z_0$, this follows by observing that both f and $1/(\cdot - z_0)^m$ are holomorphic in $D \setminus \{0\}$. For $|z - z_0| < R$, this follows since g is given by a power series!

(2) $g(z_0) = c_m \neq 0$. (Definition of m!)

(3) $f(z) = (z - z_0)^m g(z)$ for $z \neq D \setminus \{z_0\}$ for $z \in D \setminus \{z_0\}$ follows from (4.4). On the other hand if $z = z_0$, then both sides are zero. Thus for all $z \in D$, $f(z) = (z - z_0)^m g(z)$.

(4) z_0 is a zero of order m. Indeed, $c_n = f^n(z_0)/n!$ for all n's, and so the claim follows using $c_m \neq 0$, while $c_0 = c_1 = \cdots = c_{m-1} = 0$.

This completes the proof. □

Example 4.6.

(1) $n\pi$ is a zero of $\sin z$ for each $n \in \mathbb{Z}$. We know that $\sin z$ is not identically zero in any neighbourhood of $n\pi$ by just looking at the restriction of $\sin z$ to the real axis. We ask: what is the order of $n\pi$ as a zero of $\sin z$? Since $\sin' z = \cos z$ and $\cos z|_{z=n\pi} = (-1)^n \neq 0$, it follows that $n\pi$ is a zero of $\sin z$ of order 1.

(2) $\exp(z^2) - 1$ has a zero at 0 since $\exp(0^2) - 1 = 1 - 1 = 0$. What is its order? We have

$$\exp(z^2) = 1 + \frac{z^2}{1!} + \frac{z^4}{2!} + \cdots, \quad z \in \mathbb{C},$$

and so $\exp(z^2) - 1 = z^2 g(z)$ $(z \in \mathbb{C})$, where $g(z) := \frac{1}{1!} + \frac{z^2}{2!} + \cdots$. g is given by a power series that converges in $\mathbb{C}$, and so g is entire. Also, $g(0) = 1 \neq 0$. Thus the order of 0 as a zero of $\exp(z^2) - 1$ is 2. Alternately, we could have observed that

$$\frac{d}{dz}(\exp(z^2) - 1)\Big|_{z=0} = (\exp(z^2)) \cdot 2z\Big|_{z=0} = 0,$$

$$\frac{d^2}{dz^2}(\exp(z^2) - 1)\Big|_{z=0} = (\exp(z^2)) \cdot 2 + (\exp(z^2)) \cdot (2z)^2\Big|_{z=0}$$

$$= 2 \neq 0,$$

and so the order of 0 as a zero of $\exp(z^2) - 1$ is 2. ◊

Exercise 4.17. Let D be a domain, $m \in \mathbb{N}$, $R > 0$ and $z_0 \in D$. Let $f, g : D \to \mathbb{C}$ be holomorphic such that $g(z_0) \neq 0$ and whenever $|z - z_0| < R$, $f(z) = (z - z_0)^m g(z_0)$. Prove that z_0 is a zero of f of order m.

Exercise 4.18. Find the order of the zero z_0 for the function f in each case:

(1) $z_0 = i$ and $f(z) = (1 + z^2)^4$.

(2) $z_0 = 2n\pi i$, where n is an integer, and $f(z) = \exp z - 1$.

(3) $z_0 = 0$ and $f(z) = \cos z - 1 + \dfrac{1}{2}(\sin z)^2$.

Exercise 4.19. Let f be holomorphic in a disc that contains a circle γ in its interior. Suppose there is exactly one zero z_0 of order 1 of f, which lies in the interior of γ. Prove that

$$z_0 = \frac{1}{2\pi i} \int_\gamma \frac{z f'(z)}{f(z)} \, dz.$$

Exercise 4.20. Let D be a domain and f be holomorphic in D such that f has a zero of order $m > 1$ at $z_0 \in D$. Prove that the function $z \mapsto (f(z))^2$ has a zero of order $2m$ at z_0, and that f' has a zero of order $m - 1$ at z_0.

Exercise 4.21. A complex valued continuous function on a domain needn't exhibit the dichotomy of behaviour of its zeros as in Theorem 4.1 for holomorphic functions. Give an example of a function $f : \mathbb{C} \to \mathbb{C}$, such that 0 is neither an isolated zero nor is it the case that f is identically zero in a small disc around 0 centered at 0.

4.5　The Identity Theorem

In this section, we will learn the Identity Theorem, which once again highlights the rigidity of holomorphic functions. It says roughly that a nonzero holomorphic function cannot have an accumulation of zeros in its domain.

Theorem 4.5. *Let*

(1) *D be a domain,*

(2) *$f : D \to \mathbb{C}$ be a holomorphic function in D,*

(3) *$(z_n)_{n \in \mathbb{N}}$ be a sequence of distinct zeros of f which converges to $z_* \in D$.*

Then f is identically zero in D.

Proof.　First let us note that z_* is itself a zero of f because by the continuity of f, we have

$$f(z_*) = f\left(\lim_{n \to \infty} z_n \right) = \lim_{n \to \infty} f(z_n) = \lim_{n \to \infty} 0 = 0.$$

We claim that $f \equiv 0$ in *some* disc Δ with center z_* and radius $r > 0$. If not, then z_* is a zero of some order m, and $f(z) = (z - z_*)^m g(z)$ for z in Δ, and

$g(z) \neq 0$ in Δ. But then for all large n, $0 = f(z_n) = (z_n - z_*)^m g(z_n) \neq 0$, a contradiction.

Next we will show that $f \equiv 0$ in all of D. Suppose that $f(w) \neq 0$ for some $w \in D$. Then there is a path $\gamma : [0,1] \to D$ that joins z_* to w: $\gamma(0) = z_*$ and $\gamma(1) = w$. Let $S := \{t \in [0,1] : f(\gamma(\tau)) = 0 \text{ for } 0 \leq \tau \leq t\}$, and $T := \sup S$. We note that $\sup S$ exists since S is nonempty (0 belongs to S!) and S is bounded above (by 1). What is T? If we think of t as time as we travel from z_* (time $t = 0$) to w (time $t = 1$) along γ, then T is the largest time such that f has been 0 along the path covered so far. If $T = 1$, then we are done (since then by continuity, $f(w) = f(T) = 0$). So let us suppose that $T < 1$. Then $f(\gamma(T)) = 0$, by continuity. But then $f(z) = 0$ for z's in a disc around $\gamma(T)$ of radius say δ, since $\gamma(T)$ can't be an isolated zero! This implies that $f(\gamma(t)) = 0$ for t's that are bigger than T (because for all t's close enough to T, $|\gamma(t) - \gamma(T)| < r$).

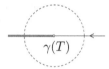

This contradicts the definition of T. Thus T can't be less than 1. This completes the proof. $\qquad\square$

Example 4.7. We had shown that for all $z \in \mathbb{C}$, $(\cos z)^2 + (\sin z)^2 = 1$ using the definitions of $\cos z$ and $\sin z$. Here is a proof using the above result. Consider $f : \mathbb{C} \to \mathbb{C}$, $f(z) = (\cos z)^2 + (\sin z)^2 - 1$, $z \in \mathbb{C}$. Then f is entire. Also, for all $x \in \mathbb{R}$, $f(x) = (\cos x)^2 + (\sin x)^2 - 1 = 0$ by Pythagoras's Theorem. So by the result above, $f \equiv 0$ in $\mathbb{C}$! $\qquad\lozenge$

The following is an immediate consequence of this result.

Corollary 4.6. (Identity Theorem) *Let*

(1) D *be a domain,*
(2) $f, g : D \to \mathbb{C}$ *be holomorphic in* D,
(3) $(z_n)_{n \in \mathbb{N}}$ *be a sequence of distinct points in* D *which converges to* $z_* \in D$, *and such that for all* $n \in \mathbb{N}$, $f(z_n) = g(z_n)$.

Then $f(z) = g(z)$ *for all* $z \in D$.

Proof. Define $h : D \to \mathbb{C}$ by $h(z) = f(z) - g(z)$, $z \in D$, and note that the z_ns are zeros of the holomorphic function h. By the result above, h

must be identically zero in D, and so the claim follows. □

Example 4.8. We know that $\exp : \mathbb{C} \to \mathbb{C}$ defined by

$$\exp z = \exp(x + iy) := e^x(\cos y + i \sin y), \quad z = x + iy \in \mathbb{C},$$

is an entire function such that $\exp x = e^x$ for $x \in \mathbb{R}$. In other words, $\exp$ is an entire extension of the usual real exponential function. Is there any other entire extension possible? We show that the answer is no! Suppose that $g : \mathbb{C} \to \mathbb{C}$ is entire and $g(x) = e^x$ for real x. But then

$$\exp x = g(x) \text{ for all } x \in \mathbb{R}.$$

In particular,

$$\exp\left(\frac{1}{n}\right) = g\left(\frac{1}{n}\right), \quad n \in \mathbb{N},$$

and $1/n \to 0 \in \mathbb{C}$. So by the Identity Theorem, $\exp z = g(z)$ for all $z \in \mathbb{C}$. So there is *only one* entire function whose restriction to $\mathbb{R}$ is e^x. This explains the naturalness of the definition of the complex exponential in Section 1.4.1. ◇

Exercise 4.22. Show, using the Identity Theorem, that for all $z_1, z_2 \in \mathbb{C}$,

$$\cos(z_1 + z_2) = (\cos z_1)(\cos z_2) - (\sin z_1)(\sin z_2),$$

by appealing to the corresponding identity when z_1, z_2 are real numbers.

Exercise 4.23. Let D be a domain and let $\mathrm{Hol}(D)$ be the set of all functions holomorphic in D. Then it is easy to check that $\mathrm{Hol}(D)$ is a commutative ring with the pointwise operations

$$(f + g)(z) = f(z) + g(z),$$
$$(f \cdot g)(z) = f(z)g(z),$$

for $z \in D$ and $f, g \in \mathrm{Hol}(D)$. (By a *commutative ring* R, we mean a set R with two laws of composition $+$ and $\cdot$ such that $(R, +)$ is an Abelian group, $\cdot$ is associative, commutative and has an identity, and the distributive law holds: for $a, b, c \in R$, $(a + b) \cdot c = a \cdot c + b \cdot c$.)

Check that $\mathrm{Hol}(D)$ is an *integral domain*, that is, a nonzero ring having no zero divisors. In other words, if $f \cdot g = 0$ for $f, g \in \mathrm{Hol}(D)$, then either $f = 0$ or $g = 0$.

If instead of $\mathrm{Hol}(D)$, we consider the set $C(D)$ of all complex-valued *continuous* functions on D, then $C(D)$ is again a commutative ring with pointwise operations. Is $C(D)$ an integral domain? (This shows that continuous functions are not as "rigid" as holomorphic functions.)

Exercise 4.24. Let f, g be holomorphic functions in a domain D. Which of the following conditions imply $f = g$ identically in D?

(1) There is a sequence $(z_n)_{n \in \mathbb{N}}$ of distinct points in D such that $f(z_n) = g(z_n)$ for all $n \in \mathbb{N}$.

(2) There is a convergent sequence $(z_n)_{n \in \mathbb{N}}$ of distinct points in D with its limit in D such that $f(z_n) = g(z_n)$ for all $n \in \mathbb{N}$.

(3) γ is a smooth path in D joining distinct points $a, b \in D$ and $f = g$ on γ.

(4) $w \in D$ is such that $f^{(n)}(w) = g^{(n)}(w)$ for all $n \geq 0$.

Exercise 4.25. Suppose that f is an entire function, and that in every power series (that is, for every $z_0 \in \mathbb{C}$) $f(z) = \sum_{n=0}^{\infty} c_n(z - z_0)^n$, at least one coefficient is 0. Prove that f is a polynomial.

4.6 The Maximum Modulus Theorem

In this section, we prove an important result, known as the Maximum Modulus Theorem which says that for a nonconstant holomorphic function $f : D \to \mathbb{C}$, $|f|$ can't have a maximizer in the domain D.

Theorem 4.6. (Maximum Modulus Theorem) *Let*

(1) *D be a domain,*
(2) *$f : D \to \mathbb{C}$ be holomorphic in D,*
(3) *$z_0 \in D$ be such that for all $z \in D$, $|f(z_0)| \geq |f(z)|$.*

Then f is constant on D.

Proof. Let $r > 0$ be such that the disc with center z_0 and radius $2r$ is contained in D. Let C_r be the circular path $C_r(t) = z_0 + r \exp(it)$, $t \in [0, 2\pi]$. Then by the Cauchy Integral Formula,

$$f(z_0) = \frac{1}{2\pi i} \int_\gamma \frac{f(z)}{z - z_0} dz = \frac{1}{2\pi i} \int_0^{2\pi} \frac{f(z_0 + r \exp(it))}{r \exp(it)} i r \exp(it) dt$$

$$= \frac{1}{2\pi} \int_0^{2\pi} f(z_0 + r \exp(it)) dt,$$

where the last expression can be viewed as the "average" of the values of f on C_r. Since $|f(z_0 + r \exp(it))| \leq |f(z_0)|$ for all t, the above yields

$$|f(z_0)| = \left| \frac{1}{2\pi} \int_0^{2\pi} f(z_0 + r\exp(it))dt \right| \leq \frac{1}{2\pi} \int_0^{2\pi} |f(z_0 + r\exp(it))|dt$$

$$\leq \frac{1}{2\pi} \int_0^{2\pi} |f(z_0)|dt = |f(z_0)|.$$

So in the above, all the inequalities $\leq$ have to be equalities, and by rearranging, we get

$$\frac{1}{2\pi} \int_0^{2\pi} \Big(\underbrace{|f(z_0)| - |f(z_0 + r\exp(it))|}_{\geq 0} \Big) dt = 0.$$

But the integrand is pointwise nonnegative, and so in light of the above, we can conclude that $|f(z_0 + r\exp(it))| = |f(z_0)|$ for all t. But by replacing r by any smaller number, the same conclusion would hold. Thus f maps the disc $\Delta := \{z \in \mathbb{C} : |z - z_0| \leq r\}$ into the circle $\{w \in \mathbb{C} : |w| = |f(z_0)|\}$. This implies by Example 2.11 that f is constant on Δ. The Identity Theorem now implies that f must be constant on the whole of D. $\qquad\square$

Example 4.9. Let $\mathbb{H} := \{z \in \mathbb{C} : \mathrm{Re}(z) \geq 0\}$ denote the right half plane, and consider $f : \mathbb{H} \to \mathbb{C}$ given by

$$f(z) = \frac{\exp(-z)}{z+1}, \quad z \in \mathbb{H}.$$

Then it can be shown that

$$\|f\|_\infty := \max_{z \in \mathbb{H}} |f(z)|$$

exists. Without worrying about the existence of this maximum, let us instead see how, assuming its existence, the Maximum Modulus Theorem enables us to calculate its value. Suppose that $z_0 \in \mathbb{H}$ is a maximizer. Then this maximizer z_0 can't have a real part which is positive, by the Maximum Modulus Theorem. So $z_0 \in i\mathbb{R}$, that is, $z_0 = iy_0$ for some $y_0 \in \mathbb{R}$. But

$$|f(iy)| = \left| \frac{\exp(-iy)}{iy+1} \right| = \frac{1}{\sqrt{y^2+1}}, \quad y \in \mathbb{R}.$$

Thus $\|f\|_\infty = \max_{z \in \mathbb{H}} |f(z)| = \max_{y \in \mathbb{R}} |f(iy)| = \max_{y \in \mathbb{R}} \frac{1}{\sqrt{y^2+1}} = \frac{1}{\sqrt{0^2+1}} = 1.$ $\Diamond$

Exercise 4.26. Let D be a domain and let $f : D \to \mathbb{C}$ be a nonconstant holomorphic map. Prove that there is no maximizer for the map $z \mapsto |f(z)|$ on D.

Exercise 4.27. (Minimum Modulus Theorem) Let D be a domain and let $f : D \to \mathbb{C}$ be holomorphic in D. Suppose that there is a $z_0 \in D$ such that for all $z \in D$, $|f(z_0)| \leq |f(z)|$. Then prove that either $f(z_0) = 0$ or f is constant on D.

Exercise 4.28. Consider the function f defined by $f(z) = z^2 - 2$. Find the maximum and minimum value of $|f(z)|$ on $\{z \in \mathbb{C} : |z| \leq 1\}$.

4.7   Laurent series

Laurent series generalize Taylor series. Indeed, while a Taylor series

$$\sum_{n=0}^{\infty} c_n (z - z_0)^n$$

has nongenative powers of the term $z - z_0$, and converges in a disc, a *Laurent series* is an expression of the type

$$\sum_{n \in \mathbb{Z}} c_n (z - z_0)^n = \cdots + c_{-1}(z - z_0)^{-1} + c_0 + c_1(z - z_0)^1 + \cdots ,$$

which has negative powers of $z - z_0$ too.

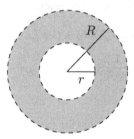

We will see that

(1) Laurent series "converge" in an annulus $\{z \in \mathbb{C} : r < |z - z_0| < R\}$ with center z_0 and gives a holomorphic function there, and

(2) conversely, if we have a holomorphic function in an annulus with center z_0 and it has singularities that lie in the "hole" inside the annulus, then the function has a Laurent series expansion in the annulus. For example, we know that for all $z \in \mathbb{C}$,

$$\exp z = 1 + \frac{z}{1!} + \frac{z^2}{2!} + \frac{z^3}{3!} + \cdots ,$$

and so for $z \neq 0$, we have the "Laurent series expansion"

$$\exp \frac{1}{z} = 1 + \frac{1}{z} + \frac{1}{2!}\frac{1}{z^2} + \frac{1}{3!}\frac{1}{3!} + \cdots .$$

Note that $\exp(1/z)$ is holomorphic in $\mathbb{C} \setminus \{0\}$, which is a degenerate annulus centered at 0 with inner radius $r = 0$ and outer radius $R = +\infty$!

Let us first define what we mean by the convergence of $\displaystyle\sum_{n \in \mathbb{Z}} c_n (z - z_0)^n$.

Definition 4.3.
The Laurent series $\sum_{n \in \mathbb{Z}} c_n(z - z_0)^n$ *converges* (for z) if

$$\sum_{n=1}^{\infty} c_{-n}(z - z_0)^{-n} \text{ converges and } \sum_{n=0}^{\infty} c_n(z - z_0)^n \text{ converges.}$$

If $\sum_{n \in \mathbb{Z}} c_n(z - z_0)^n$ converges, then we write

$$\sum_{n \in \mathbb{Z}} c_n(z - z_0)^n = \sum_{n=1}^{\infty} c_{-n}(z - z_0)^{-n} + \sum_{n=0}^{\infty} c_n(z - z_0)^n,$$

and call this the *sum* of the Laurent series.

Example 4.10. For what $z \in \mathbb{C}$ does the Laurent series

$$\cdots + \frac{1}{8z^3} + \frac{1}{4z^2} + \frac{1}{2z} + 1 + z + z^2 + z^3 + \cdots$$

converge? We have:

(1) $1 + z + z^2 + z^3 + \cdots$ converges for $|z| < 1$, and it diverges for $|z| > 1$.

(2) $\dfrac{1}{2z} + \dfrac{1}{4z^3} + \dfrac{1}{8z^5} + \cdots$ converges for $\left|\dfrac{1}{2z}\right| < 1$ and diverges for $\left|\dfrac{1}{2z}\right| > 1$,

that is, it converges for $|z| > 1/2$ and diverges for $|z| < 1/2$.

Hence the given Laurent series converges when $|z| < 1$ and $|z| > 1/2$, that is, it converges inside the annulus $\{z \in \mathbb{C} : 1/2 < |z| < 1\}$, and it diverges when $|z| > 1$ or when $|z| < 1/2$. $\diamond$

For what z does $\sum_{n \in \mathbb{Z}} c_n(z - z_0)^n$ converge?

(1) From Theorem 4.1, for $\sum_{n=0}^{\infty} c_n(z - z_0)^n$, there is some R such that it converges for $|z - z_0| < R$ and diverges for $|z - z_0| > R$.

(2) What about the series $\sum_{n=1}^{\infty} c_{-n}(z - z_0)^{-n}$? The power series $\sum_{n=1}^{\infty} c_{-n}w^n$ also converges for $|w| < \widetilde{R}$ and diverges for $|w| > \widetilde{R}$. Set $w := (z - z_0)^{-1}$. Then $\sum_{n=1}^{\infty} c_{-n}(z - z_0)^{-n}$ converges when $1/|z - z_0| < \widetilde{R}$, that is for $|z - z_0| > 1/\widetilde{R} =: r$, and diverges for $|z - z_0| < r$.

Hence the Laurent series converges in the annulus $\{z \in \mathbb{C} : r < |z - z_0| < R\}$ and diverges if either $|z - z_0| < r$ or $|z - z_0| > R$.

Is it holomorphic in the annulus where it converges?

(1) $z \mapsto \sum_{n=0}^{\infty} c_n (z - z_0)^n$ is holomorphic in $\{z \in \mathbb{C} : |z - z_0| < R\}$ and so in particular, also in $\{z \in \mathbb{C} : r < |z - z_0| < R\}$.

(2) The map

$$w \overset{g}{\longmapsto} \sum_{n=1}^{\infty} c_{-n} w^n$$

is holomorphic in $\{w \in \mathbb{C} : |w| < \widetilde{R}\}$. Also we see that the mapping $z \overset{f}{\longmapsto} (z - z_0)^{-1} : \mathbb{C} \setminus \{z_0\} \to \mathbb{C}$ is holomorphic. So their composition $g \circ f$ is holomorphic in $\{z \in \mathbb{C} : |z - z_0| > r\}$, that is,

$$z \overset{g \circ f}{\longmapsto} \sum_{n=1}^{\infty} c_{-n} (z - z_0)^{-n}$$

is holomorphic in $\{z \in \mathbb{C} : r < |z - z_0|\}$, and so also in particular in the annulus $\{z \in \mathbb{C} : r < |z - z_0| < R\}$.

Hence the sum $z \mapsto \sum_{n=0}^{\infty} c_n (z - z_0)^n + \sum_{n=1}^{\infty} c_{-n} (z - z_0)^{-n} = \sum_{n \in \mathbb{Z}} c_n (z - z_0)^n$ is holomorphic in the annulus $\{z \in \mathbb{C} : r < |z - z_0| < R\}$.

Summarizing, we have learnt that any Laurent series

$$\sum_{n \in \mathbb{Z}} c_n (z - z_0)^n$$

converges in an annulus[2] $\{z \in \mathbb{C} : r < |z - z_0| < R\}$ for some r, R, and the map

$$z \mapsto \sum_{n \in \mathbb{Z}} c_n (z - z_0)^n$$

is holomorphic in $\{z \in \mathbb{C} : r < |z - z_0| < R\}$.

That conversely, a function holomorphic in an annulus has a Laurent series expansion is the content of the following theorem.

---

[2]which may be empty!

Theorem 4.7. *If f is holomorphic in $\mathbb{A} := \{z \in \mathbb{C} : r < |z - z_0| < R\}$, then*

$$f(z) = \sum_{n \in \mathbb{Z}} c_n (z - z_0)^n \quad \text{for } z \in \mathbb{A}, \tag{4.5}$$

where

(1) $c_n = \dfrac{1}{2\pi i} \displaystyle\int_C \dfrac{f(\zeta)}{(\zeta - z_0)^{n+1}} d\zeta$,

(2) *C is the circular path given by $C(t) = z_0 + \rho \exp(it)$, $t \in [0, 2\pi]$,*

(3) *ρ is any number such that $r < \rho < R$.*

Moreover, the coefficients are unique in (4.5).

Example 4.11. Define $f : \mathbb{C} \setminus \{0\} \to \mathbb{C}$ by $f(z) = z^3 \exp(1/z)$, $z \neq 0$. Then f is holomorphic in $\mathbb{A} := \{z \in \mathbb{C} : 0 < |z| < +\infty\}$. So by the above result, f must have a Laurent expansion

$$\sum_{n \in \mathbb{Z}} c_n z^n.$$

In this case we can find the coefficients just by inspection: since

$$\exp \frac{1}{z} = 1 + \frac{1}{z} + \frac{1}{2! z^2} + \cdots, \quad z \neq 0,$$

we have $f(z) = z^3 \exp \dfrac{1}{z} = z^3 + z^2 + \dfrac{z}{2!} + \dfrac{1}{3!} + \dfrac{1}{4! z} + \cdots$, for $z \neq 0$. Consequently,

$$\cdots, \ c_{-1} = \frac{1}{4!}, \ c_0 = \frac{1}{3!}, \ c_1 = \frac{1}{2!}, \ c_2 = 1, \ c_3 = 1,$$

and $c_n = 0$ for $n \geq 4$. $\Diamond$

Proof. (of Theorem 4.7.) (Existence.) Fix $z \in \mathbb{A}$. Choose $\widetilde{r}$ and $\widetilde{R}$ such that $r < \widetilde{r} < |z - z_0| < \widetilde{R} < R$. Let γ_1 and γ_2 be the circular paths

$$\gamma_1(t) = z_0 + \widetilde{r} \exp(it),$$
$$\gamma_2(t) = z_0 + \widetilde{R} \exp(it),$$

for $t \in [\theta, 2\pi + \theta]$, and $\theta := \operatorname{Arg}(z) + \pi/2$. Let $\gamma_3 : [\widetilde{r}, \widetilde{R}] \to \mathbb{A}$ be the path

$$\gamma_3(t) = ti \frac{z - z_0}{|z - z_0|}.$$

(This is just a straight line path joining γ_1 and γ_2, and the peculiar multiplication by i produces a rotation by $90°$, ensuring that this path avoids z!) See Figure 4.2.

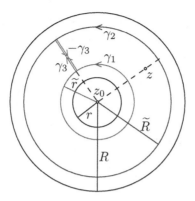

Fig. 4.2   Laurent series.

Clearly the path $\gamma := \gamma_2 - \gamma_3 - \gamma_1 + \gamma_3$ is $\mathbb{A} \setminus \{z\}$-homotopic to a small circle C_δ centered at z. Also, $\dfrac{f(\cdot)}{\cdot - z}$ is holomorphic in $\mathbb{A} \setminus \{z\}$, and so

$$\int_\gamma \frac{f(\zeta)}{\zeta - z}d\zeta = \int_{C_\delta} \frac{f(\zeta)}{\zeta - z}d\zeta = f(z) \cdot 2\pi i,$$

where the first equality follows from the Cauchy Integral Theorem, and the second equality follows from the Cauchy Integral Formula. Thus since the contour integral along γ_3 cancels with that along $-\gamma_3$, we obtain

$$f(z) = \frac{1}{2\pi i} \int_\gamma \frac{f(\zeta)}{\zeta - z}d\zeta = \frac{1}{2\pi i} \left[\int_{\gamma_2} - \int_{\gamma_3} - \int_{\gamma_1} + \int_{\gamma_3} \right] \frac{f(\zeta)}{\zeta - z}d\zeta$$

$$= \underbrace{\frac{1}{2\pi i} \int_{\gamma_2} \frac{f(\zeta)}{\zeta - z}d\zeta}_{(\text{I})} - \underbrace{\frac{1}{2\pi i} \int_{\gamma_1} \frac{f(\zeta)}{\zeta - z}d\zeta}_{(\text{II})}.$$

We will see below that the integral (I) gives the term

$$\sum_{n=0}^\infty c_n(z - z_0)^n,$$

while the integral (II) gives the term

$$\sum_{n=1}^\infty c_{-n}(z - z_0)^{-n},$$

and these put together we will yield the desired Laurent series expansion of f.

Step 1. In this step we will show that $\dfrac{1}{2\pi i} \displaystyle\int_{\gamma_2} \dfrac{f(\zeta)}{\zeta - z} d\zeta = \displaystyle\sum_{n=0}^{\infty} c_n (z - z_0)^n$.

We have for $\zeta \in \gamma_2$ that

$$\frac{f(\zeta)}{\zeta - z} = \frac{f(\zeta)}{\zeta - z_0 + z_0 - z} = \frac{f(\zeta)}{(\zeta - z_0)\left(1 - \dfrac{z - z_0}{\zeta - z_0}\right)} = \frac{f(\zeta)}{(\zeta - z_0)(1 - w)},$$

where $w := \dfrac{z - z_0}{\zeta - z_0}$. We have $|w| = \dfrac{|z - z_0|}{|\zeta - z_0|} = \dfrac{|z - z_0|}{\widetilde{R}} < 1$, and so

$$\frac{1}{1 - w} = 1 + w + w^2 + w^3 + \cdots + w^{n-1} + \frac{w^n}{1 - w}.$$

Using this, we obtain

$$\frac{f(\zeta)}{\zeta - z} = \frac{f(\zeta)}{\zeta - z_0}\left(1 + w + \cdots + w^n + \frac{w^n}{1 - w}\right)$$

$$= \frac{f(\zeta)}{\zeta - z_0} + \cdots + \frac{f(\zeta)}{(\zeta - z_0)^n}(z - z_0)^{n-1} + \frac{f(\zeta)(z - z_0)^n}{(\zeta - z_0)^n(\zeta - z)}.$$

Thus

$$\frac{1}{2\pi i} \int_{\gamma_2} \frac{f(\zeta)}{\zeta - z} d\zeta = \frac{1}{2\pi i} \int_{\gamma_2} \frac{f(\zeta)}{\zeta - z_0} d\zeta + \cdots + \frac{1}{2\pi i} \int_{\gamma_2} \frac{f(\zeta)}{(\zeta - z_0)^n} d\zeta \cdot (z - z_0)^{n-1}$$

$$+ \frac{1}{2\pi i} \int_{\gamma_2} \frac{f(\zeta)}{(\zeta - z_0)^n(\zeta - z)} d\zeta \cdot (z - z_0)^n$$

$$= c_0 + c_1(z - z_0) + \cdots + c_{n-1}(z - z_0)^{n-1} + R_n(z),$$

where

$$R_n(z) := \frac{1}{2\pi i} \int_{\gamma_2} \frac{f(\zeta)(z - z_0)^n}{(\zeta - z_0)^n(\zeta - z)} d\zeta.$$

Here we have used the fact that γ_2 is $\mathbb{A}$-homotopic to any circle C with center z_0 and radius ρ where $r < \rho < R$, and so by Theorem 3.4,

$$\int_{\gamma_2} \frac{f(\zeta)}{(\zeta - z_0)^k} d\zeta = \int_{C} \frac{f(\zeta)}{(\zeta - z_0)^k} d\zeta = c_k$$

for $k = 1, \ldots, n - 1$. We would be done and obtain

$$\frac{1}{2\pi i} \int_{\gamma_2} \frac{f(\zeta)}{\zeta - z} d\zeta = \sum_{n=0}^{\infty} c_n(z - z_0)^n$$

if we show that $\displaystyle\lim_{n \to \infty} R_n = 0$. There is an $M > 0$ such that for all $\zeta \in \gamma_2$, $|f(\zeta)| < M$. Moreover, $|\zeta - z_0| = \widetilde{R}$ and

$$|\zeta - z| = |\zeta - z_0 - (z - z_0)| \geq |\zeta - z_0| - |z - z_0| = \widetilde{R} - |z - z_0|,$$

and so

$$|R_n(z)| \leq \left(\frac{|z - z_0|}{\widetilde{R}}\right)^n \frac{M}{\widetilde{R} - |z - z_0|} \xrightarrow{n \to \infty} 0.$$

Thus $c_0 + c_1(z - z_0) + c_2(z - z_0)^2 + c_3(z - z_0)^3 + \cdots = \dfrac{1}{2\pi i} \displaystyle\int_{\gamma_2} \dfrac{f(\zeta)}{\zeta - z} d\zeta.$

Step 2. We will show that $-\dfrac{1}{2\pi i} \displaystyle\int_{\gamma_1} \dfrac{f(\zeta)}{\zeta - z} d\zeta = \displaystyle\sum_{n=1}^{\infty} c_{-n}(z - z_0)^{-n}.$

We have

$$-\frac{1}{2\pi i} \int_{\gamma_1} \frac{f(\zeta)}{\zeta - z} d\zeta = \frac{1}{2\pi i} \int_{\gamma_1} \frac{f(\zeta)}{(z - z_0) - (\zeta - z_0)} d\zeta$$

$$= \frac{1}{2\pi i} \int_{\gamma_1} \frac{f(\zeta)}{(z - z_0)\left(1 - \dfrac{\zeta - z_0}{z - z_0}\right)} d\zeta$$

Set $w := \dfrac{\zeta - z_0}{z - z_0}$. Then $|w| = \dfrac{\widetilde{r}}{|z - z_0|} < 1$. Thus

$$\frac{1}{1 - \dfrac{\zeta - z_0}{z - z_0}} = \frac{1}{1 - w} = 1 + w + w^2 + w^3 + \cdots + w^{n-1} + \frac{w^n}{1 - w},$$

and so

$$-\frac{1}{2\pi i} \int_{\gamma_1} \frac{f(\zeta)}{\zeta - z} d\zeta$$

$$= \frac{1}{2\pi i} \int_{\gamma_1} f(\zeta) \left(\frac{1}{z - z_0} + \cdots + \frac{(\zeta - z_0)^{n-1}}{(z - z_0)^n} + \frac{(\zeta - z_0)^n}{(z - z_0)^n(z - \zeta)}\right) d\zeta$$

$$= \frac{1}{2\pi i} \int_{\gamma_1} f(\zeta) d\zeta \cdot \frac{1}{z - z_0} + \cdots + \frac{1}{2\pi i} \int_{\gamma_1} \frac{f(\zeta)}{(\zeta - z_0)^{-n+1}} d\zeta \cdot \frac{1}{(z - z_0)^n}$$

$$+ \frac{1}{2\pi i} \int_{\gamma_1} \frac{f(\zeta)(\zeta - z_0)^n}{(z - z_0)^n(z - \zeta)} d\zeta$$

$$= c_{-1}(z - z_0)^{-1} + \cdots + c_{-(n-1)}(z - z_0)^{-(n-1)} + \widetilde{R}_n(z),$$

where

$$\widetilde{R}_n(z) := \frac{1}{2\pi i} \int_{\gamma_1} \frac{f(\zeta)(\zeta - z_0)^n}{(z - z_0)^n(z - \zeta)} d\zeta.$$

Here we have used the fact that γ_1 is $\mathbb{A}$-homotopic to any circle C with center z_0 and radius ρ where $r < \rho < R$, and so by Theorem 3.4,

$$\int_{\gamma_1} \frac{f(\zeta)}{(\zeta - z_0)^k} d\zeta = \int_C \frac{f(\zeta)}{(\zeta - z_0)^k} d\zeta = c_k$$

for $k = -1, \ldots, -(n-1)$. There is an $M > 0$ such that for all $\zeta \in \gamma_1$, $|f(\zeta)| < M$. Moreover, $|\zeta - z_0| = \widetilde{r}$ and

$$|z - \zeta| = |(z - z_0) - (\zeta - z_0)| \geq |z - z_0| - |\zeta - z_0| = |z - z_0| - \widetilde{r},$$

and so

$$|\widetilde{R}_n(z)| \leq \left(\frac{\widetilde{r}}{|z - z_0|}\right)^n \frac{M}{|z - z_0| - \widetilde{r}} \overset{n \to \infty}{\longrightarrow} 0.$$

Thus $c_{-1}(z - z_0)^{-1} + \cdots + c_{-(n-1)}(z - z_0)^{-(n-1)} + \cdots = -\dfrac{1}{2\pi i} \displaystyle\int_{\gamma_1} \dfrac{f(\zeta)}{\zeta - z} d\zeta.$

This completes the proof of the part on the existence of the Laurent expansion.

Uniqueness of coefficients. Cauchy's Integral Formula allows us to show that the Laurent expansion is unique, that is, if

$$f(z) = \sum_{n \in \mathbb{Z}} \widetilde{c}_n (z - z_0)^n \text{ for } r < |z - z_0| < R,$$

then for all n, $\widetilde{c}_n = c_n$. If $n \neq 1$, then $(z - z_0)^n = \dfrac{d}{dz}\left(\dfrac{(z - z_0)^{n+1}}{n+1}\right)$. So

$$\int_C (z - z_0)^n dz = 0 \quad (n \neq 1),$$

where C is given by $C(t) = z_0 + \rho \exp(it)$, $t \in [0, 2\pi]$. By a direct calculation,

$$\int_C \frac{1}{z - z_0} dz = \int_0^{2\pi} \frac{1}{r \exp(it)} ir \exp(it) dt = 2\pi i.$$

Hence, if term-by-term integration is justified in the annulus, we would have

$$\int_C (z - z_0)^{-m-1} \sum_{n \in \mathbb{Z}} \widetilde{c}_n (z - z_0)^n dz = \sum_{n \in \mathbb{Z}} \widetilde{c}_n \int_C (z - z_0)^{n-m-1} dz = 2\pi i \widetilde{c}_m,$$

and the claim about the uniqueness of coefficients would be proved. This term-by-term integration can be justified as follows. We have

$$\sum_{n \in \mathbb{Z}} \widetilde{c}_n (z - z_0)^{n-m-1} = \left(\cdots + \frac{\widetilde{c}_{m-2}}{(z - z_0)^3} + \frac{\widetilde{c}_{m-1}}{(z - z_0)^2}\right)$$

$$+ \frac{\widetilde{c}_m}{z - z_0} + (\widetilde{c}_{m+1} + \widetilde{c}_{m+2}(z - z_0) + \cdots)$$

$$= f_1(z) + \frac{\widetilde{c}_m}{z - z_0} + f_2(z).$$

We need only show that f_1, f_2 have a primitive in the annulus and then

$$\int_C \sum_{n \in \mathbb{Z}} \widetilde{c}_n (z - z_0)^{n-m-1} dz = \int_C \left(f_1(z) + \frac{\widetilde{c}_m}{z - z_0} + f_2(z) \right) dz$$

$$= 0 + 2\pi i \widetilde{c}_m + 0 = 2\pi i \widetilde{c}_m,$$

as required. But $f_2(z) = \sum_{n=1}^{\infty} \widetilde{c}_{m+n} (z - z_0)^{n-1}$ for $|z - z_0| < R$, and so if

$$F_2(z) := \sum_{n=1}^{\infty} \frac{\widetilde{c}_{m+n}}{n} (z - z_0)^n \quad \text{for } |z - z_0| < R,$$

then $\dfrac{d}{dz} F_2(z) = f_2(z)$, and so F_2 is a primitive of f_2. For f_1, we have

$$f_1(z) = \sum_{n=1}^{\infty} \frac{\widetilde{c}_{m-n}}{(z - z_0)^{n+1}} = \sum_{n=1}^{\infty} \widetilde{c}_{m-n} w^{n+1},$$

where $w = \dfrac{1}{z - z_0}$, and this is valid for $R > |z - z_0| > r$. So

$$\sum_{n=1}^{\infty} \widetilde{c}_{m-n} w^{n+1}$$

converges for $|w| < \dfrac{1}{r}$. If we set

$$G(w) = -\sum_{n=1}^{\infty} \frac{\widetilde{c}_{m-n}}{n} w^n \quad \text{for } |w| < \frac{1}{r},$$

then $\dfrac{d}{dz} G(w) = -\sum_{n=1}^{\infty} \widetilde{c}_{m-n} w^{n-1}$. Hence if define F_1 by

$$F_1(z) = G\left(\frac{1}{z - z_0} \right) = \sum_{n=1}^{\infty} \frac{\widetilde{c}_{m-n}}{n} (z - z_0)^{-n} \quad \text{for } z \in \mathbb{A},$$

then

$$\frac{d}{dz} F_1(z) = \left(G'\left(\frac{1}{z - z_0} \right) \right) \left(-\frac{1}{(z - z_0)^2} \right)$$

$$= (z - z_0)^{-2} \sum_{n=1}^{\infty} \widetilde{c}_{m-n} (z - z_0)^{-n+1} = f_1(z).$$

$\square$

Note that the uniqueness of coefficients is valid only if we consider a particular fixed annulus. It can happen that the same function has different Laurent expansions, but valid in *different* annuli, as shown in the following example.

Example 4.12. Consider f defined by $f(z) = \dfrac{1}{z(z-1)}$, $z \in \mathbb{C} \setminus \{0,1\}$.

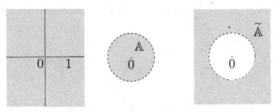

Fig. 4.3 $\mathbb{C} \setminus \{0,1\}$, $\mathbb{A} := \{z \in \mathbb{C} : 0 < |z| < 1\}$ and $\widetilde{\mathbb{A}} := \{z \in \mathbb{C} : 1 < |z| < +\infty\}$.

f is holomorphic in the annulus $\mathbb{A} := \{z \in \mathbb{C} : 0 < |z| < 1\}$. What is its Laurent expansion there? Since $|z| < 1$, we have

$$f(z) = \frac{1}{z(z-1)} = -\frac{1}{z}\left(1 + z + z^2 + z^3 + \cdots\right) = -\frac{1}{z} - 1 - z - z^2 - z^3 - \cdots,$$

and so the Laurent series coefficients are given by

$$c_{-2} = c_{-3} = \cdots = 0,$$
$$c_{-1} = c_0 = c_1 = \cdots = -1.$$

But f is also holomorphic in the annulus $\widetilde{\mathbb{A}} := \{z \in \mathbb{C} : 1 < |z|\}$, with inner radius 1, and outer radius infinite. So f has a Laurent series expansion in $\widetilde{A}$ too! Since $|z| > 1$, we have

$$f(z) = \frac{1}{z(z-1)} = \frac{1}{z^2(1 - 1/z)} = \frac{1}{z^2}\left(1 + \frac{1}{z} + \frac{1}{z^2} + \frac{1}{z^3} + \cdots\right)$$
$$= \frac{1}{z^2} + \frac{1}{z^3} + \cdots.$$

Thus now the Laurent series coefficients are given by

$$\widetilde{c}_{-2} = \widetilde{c}_{-3} = \cdots = 1,$$
$$c_{-1} = c_0 = c_1 = \cdots = 0.$$

So we notice that the coefficients are different, but this is not surprising, since the annuli we considered for the Laurent expansions were different too. $\Diamond$

Exercise 4.29. Reconsider Example 4.12, and now find Laurent expansions for f also in the annuli $\mathbb{A}_1 := \{z \in \mathbb{C} : 0 < |z-1| < 1\}$ and $\widetilde{\mathbb{A}}_1 := \{z \in \mathbb{C} : 1 < |z-1|\}$.

4.8 Classification of singularities

If we look at the three functions

$$\frac{\sin z}{z}, \quad \frac{1}{z^3}, \quad \exp\frac{1}{z},$$

then we notice that each of them is not defined at 0, and refer to 0 as a "singularity" of these functions, because the function is not defined there. But we will see that each of these functions behave very differently near their common singularity. In other words, the "nature of the singularity" differs in each case. We will explain precisely how the behaviour is different in each case, and this is what we mean by classification of singularities. Moreover, we will learn two results, which will allow us to find out the type of singularity at hand. Of these two characterization results for singularities, one result will be in terms of limits, while the other will be in terms of what happens with Laurent coefficients. We first give the following definition.

Definition 4.4. Let f be a complex valued function which is not defined at a point z_0, and suppose that it is holomorphic in some punctured disc $\{z \in \mathbb{C} : 0 < |z - z_0| < R\}$ centered at z_0 with some radius $R > 0$. Then we call z_0 an *isolated singularity* of f.

Example 4.13. For example, each of the functions

$$\frac{\sin z}{z}, \quad \frac{1}{z^3}, \quad \exp\frac{1}{z},$$

has an isolated singularity at 0. On the other hand, f given by

$$f(z) := \frac{1}{\sin(1/z)}$$

has a singularity at 0, but it is not an isolated singularity. (At $z = 1/n\pi$, $n \in \mathbb{Z}$, the function f is not defined.) $\Diamond$

Definition 4.5. An isolated singularity z_0 of f is called

(1) a *removable singularity of f* if there is a function F, holomorphic in the disc $\{z \in \mathbb{C} : |z - z_0| < R\}$ such that $F = f$ in the punctured disc $\{z \in \mathbb{C} : 0 < |z - z_0| < R\}$.
(2) a *pole of f* if $\lim_{z \to z_0} |f(z)| = +\infty$, that is,
    for all $M > 0$ there is a $\delta > 0$ such that whenever $0 < |z - z_0| < \delta$, $|f(z)| > M$.
(3) an *essential singularity of f* if z_0 is neither removable nor a pole.

Example 4.14.

(1) The function f given by $f(z) = \dfrac{\sin z}{z}$ has a removable singularity at 0, since for $z \neq 0$, we have

$$\frac{\sin z}{z} = \frac{1}{z}\left(z - \frac{z^3}{3!} + \frac{z^5}{5!} - + \cdots \right) = \sum_{n=0}^{\infty} \frac{(-1)^n}{(2n+1)!} z^{2n},$$

and the right hand side, being a power series with an infinite radius of convergence (why?) defines an entire function F. Since this entire function F coincides with the given function f in the punctured plane $\mathbb{C} \setminus \{0\}$, it follows that f has a removable singularity at 0.

(2) The function $\dfrac{1}{z^3}$ has a pole at 0, since $\displaystyle\lim_{z \to 0} \frac{1}{|z|^3} = +\infty$.

(3) The function $\exp \dfrac{1}{z}$ has an essential singularity at 0. Indeed,

    (a) 0 is not a removable singularity, because for example $\displaystyle\lim_{x \searrow 0} e^{\frac{1}{x}} = +\infty$.

    (b) 0 is also not a pole, since $\displaystyle\lim_{x \nearrow 0} e^{\frac{1}{x}} = 0$, and so it can't be that
$$\lim_{z \to 0} |f(z)| = +\infty. \hspace{3cm} \Diamond$$

We will now learn our first characterization result for singularities, in terms of limiting behaviour.

Theorem 4.8. (Classification of singularities via limits) *Suppose z_0 is an isolated singularity of f. Then*

z_0 *is removable*	$\Leftrightarrow$	$\displaystyle\lim_{z \to z_0} (z - z_0)f(z) = 0.$
z_0 *is a pole*	$\Leftrightarrow$	$(a) \neg \left(\displaystyle\lim_{z \to z_0} (z - z_0)f(z) = 0 \right)$ *and* $(b)\ \exists n \in \mathbb{N}$ *such that* $\displaystyle\lim_{z \to z_0} (z - z_0)^{n+1}f(z) = 0.$ (*The smallest such n is called* *the* order *of the pole z_0 of f.*)
z_0 *is essential*	$\Leftrightarrow$	$\forall n \in \mathbb{N}\ \neg \left(\displaystyle\lim_{z \to z_0} (z - z_0)^n f(z) = 0 \right).$

Proof. (1) $\boxed{z_0 \text{ removable} \Rightarrow \displaystyle\lim_{z \to z_0} (z - z_0)f(z) = 0}$.

Let z_0 be removable, and let F be holomorphic in

$$D(z_0, R) := \{z \in \mathbb{C} : |z - z_0| < R\}$$

such that $F = f$ for $0 < |z - z_0| < R$. Then using the fact that F is continuous at z_0, we obtain

$$\lim_{z \to z_0} (z - z_0)F(z) = \lim_{z \to z_0} (z - z_0)F(z) = 0 \cdot F(z_0) = 0.$$

Hence for every $\epsilon > 0$ there exists a $\delta > 0$ such that for $0 < |z - z_0| < \delta$, there holds that

$$|(z - z_0)f(z) - 0| = |(z - z_0)F(z) - 0| < \epsilon,$$

and so

$$\lim_{z \to z_0} (z - z_0)f(z) = 0.$$

Next we will show that $\boxed{\lim_{z \to z_0} (z - z_0)f(z) = 0 \Rightarrow z_0 \text{ is removable}}$.

Suppose that

$$\lim_{z \to z_0} (z - z_0)f(z) = 0,$$

and that f is holomorphic in the punctured disc $\{z \in \mathbb{C} : 0 < |z - z_0| < R\}$. Then f has a Laurent expansion there:

$$f(z) = \sum_{n \in \mathbb{Z}} c_n(z - z_0)^n, \quad 0 < |z - z_0| < R,$$

where the coefficients c_n's are given by contour integrals along any circle C_r with center z_0 in the punctured disc and $r < R$:

$$c_n = \frac{1}{2\pi i} \int_{C_r} \frac{f(z)}{(z - z_0)^{n+1}} dz,$$

We show that $c_{-n} = 0$ for $n \in \mathbb{N}$. Given $\epsilon > 0$, take r small enough so that $|(z - z_0)f(z)| < \epsilon$ on C_r. Then we have for $n \in \mathbb{N}$ that

$$|c_{-n}| = \left| \frac{1}{2\pi i} \int_{C_r} \frac{f(z)}{(z - z_0)^{-n+1}} dz \right| = \left| \frac{1}{2\pi i} \int_{C_r} \frac{(z - z_0)f(z)}{(z - z_0)^{-n+2}} dz \right|$$

$$\leq \frac{1}{2\pi} \frac{\epsilon}{r^{-n+2}} 2\pi r = \epsilon r^{n-1} \leq \epsilon R^{n-1}.$$

Since the choice of $\epsilon > 0$ was arbitrary, it follows that $c_{-n} = 0$ for all $n \in \mathbb{N}$. Consequently, with

$$F(z) := \sum_{n=0}^{\infty} c_n(z - z_0)^n,$$

we see that F is holomorphic in $\{z \in \mathbb{C} : |z - z_0| < R\}$, and $F = f$ in the punctured disc $\{z \in \mathbb{C} : 0 < |z - z_0| < R\}$.

$$(2) \quad z_0 \text{ is a pole} \Rightarrow \begin{cases} \neg\left(\lim_{z \to z_0} (z - z_0)f(z) = 0 \right) \text{ and} \\ \exists n \in \mathbb{N} \text{ such that } \lim_{z \to z_0} (z - z_0)^{n+1} f(z) = 0. \end{cases}$$

Suppose that z_0 is a pole of f. Then z_0 is not removable (why?), and so by the first part, it follows that

$$\neg\left(\lim_{z \to z_0} (z - z_0)f(z) = 0 \right).$$

(Here $\neg$ is the symbol for negation, to be read as "it is not the case that".)

There is some $R > 0$ such that $|f(z)| > 1$ in the punctured disc

$$D := \{z \in \mathbb{C} : 0 < |z - z_0| < R\}.$$

Define g in this punctured disc D by

$$g(z) = \frac{1}{f(z)}.$$

Since z_0 is a pole of f, it follows that $\lim_{z \to z_0} g(z) = 0$. In particular, also

$$\lim_{z \to z_0} (z - z_0)g(z) = 0,$$

and so by the first part above, g has a holomorphic extension G to

$$\{z \in \mathbb{C} : |z - z_0| < R\}.$$

Also,

$$G(z_0) = \lim_{z \to z_0} G(z) = \lim_{z \to z_0} g(z) = 0.$$

So z_0 is a zero of G, and since G is not identically zero in a neighbourhood of z_0, it follows from the result on the classification of zeros that z_0 has some order $n \in \mathbb{N}$, and there is a holomorphic function H defined for $|z - z_0| < R$ such that $H(z_0) \neq 0$ and

$$G(z) = (z - z_0)^n H(z).$$

In particular, for $0 < |z - z_0| < R$, we have

$$f(z) = \frac{1}{g(z)} = \frac{1}{G(z)} = \frac{1}{(z - z_0)^n H(z)}.$$

Hence

$$\lim_{z \to z_0} (z - z_0)^{n+1} f(z) = \lim_{z \to z_0} \frac{(z - z_0)}{H(z)} = \frac{0}{H(z_0)} = 0.$$

Done!

$$\left.\begin{array}{l} \neg\left(\lim_{z \to z_0} (z - z_0)f(z) = 0\right) \text{ and} \\ \exists n \in \mathbb{N} \text{ such that } \lim_{z \to z_0} (z - z_0)^{n+1}f(z) = 0. \end{array}\right\} \Rightarrow z_0 \text{ is a pole.}$$

Choose the smallest such n, call it n_*: thus

$$\lim_{z \to z_0} (z - z_0)^{n_*+1} f(z) = 0,$$

$$\neg\left(\lim_{z \to z_0} (z - z_0)^{n_*} f(z) = 0\right).$$

So $(z - z_0)^{n_*} f(z)$ has a removable singularity at $z = z_0$. Thus there exists an F, holomorphic in $\{z \in \mathbb{C} : |z - z_0| < R\}$ such that for $0 < |z - z_0| < R$,

$$(z - z_0)^{n_*} f(z) = F(z).$$

Note that

$$F(z_0) = \lim_{z \to z_0} F(z) = \lim_{z \to z_0} (z - z_0)^{n_*} f(z) \neq 0$$

(owing to our choice of n_*.) From

$$f(z) = \frac{F(z)}{(z - z_0)^{n_*}}$$

for $0 < |z - z_0| < R$, we obtain

$$\lim_{z \to z_0} |f(z)| = \lim_{z \to z_0} \frac{|F(z)|}{|z - z_0|^{n_*}} = \underbrace{|F(z_0)|}_{\neq 0} \cdot \lim_{z \to z_0} \frac{1}{|z - z_0|^{n_*}} = +\infty.$$

So z_0 is a pole of f.

(3) This follows easily from the first two parts. Indeed, if z_0 is an essential singularity of f, then it is not removable, and so

$$\neg\left(\lim_{z \to z_0} (z - z_0)f(z) = 0\right),$$

and so we get the desired claim at least for $n = 1$. But since z_0 isn't a pole, by (2) we also get the desired conclusion for all other n's as well.

Conversely, if

$$\text{for all } n \in \mathbb{N}, \ \neg\left(\lim_{z \to z_0} (z - z_0)^n f(z) = 0\right),$$

then in particular, from the $n = 1$ case, we get by (1) that z_0 can't be a removable singularity. And from the cases for all other $n \geq 2$, we can conclude, in light of (2), that z_0 can't be a pole either. This z_0 must be an essential singularity of f. $\qquad\square$

Let us reconsider Example 4.14.

Example 4.15.

(1) We have $\lim\limits_{z \to 0} z \cdot \dfrac{\sin z}{z} = \lim\limits_{z \to 0} \sin z = \sin 0 = 0$, and so 0 is a removable singularity of $\dfrac{\sin z}{z}$.

(2) As $\neg\left(\lim\limits_{z \to 0} z \cdot \dfrac{1}{z^3} = 0 \right)$ and $\lim\limits_{z \to 0} z^4 \cdot \dfrac{1}{z^3} = \lim\limits_{z \to 0} z = 0$, 0 is a pole of $\dfrac{1}{z^3}$, and moreover the order is 3 since $\neg\left(\lim\limits_{z \to 0} z^2 \cdot \dfrac{1}{z^3} = 0 \right)$ and $\neg\left(\lim\limits_{z \to 0} z^3 \cdot \dfrac{1}{z^3} = 0 \right)$.

(3) For $x > 0$, $e^{\frac{1}{x}} = 1 + \dfrac{1}{1!x} + \dfrac{1}{2!x^2} + \dfrac{1}{3!x^3} + \cdots > \dfrac{1}{n!x^n}$. Thus for all $n \in \mathbb{N}$, $x^n e^{\frac{1}{x}} > \dfrac{1}{n!}$ and

$$\neg\left(\lim_{x \searrow 0} x^n e^{\frac{1}{x}} = 0 \right).$$

So for all $n \in \mathbb{N}$,

$$\neg\left(\lim_{z \to 0} z^n \exp \frac{1}{z} = 0 \right).$$

Thus 0 is an essential singularity of $\exp \dfrac{1}{z}$. $\Diamond$

We will now learn our second characterization result for singularities, in terms of the Laurent series coefficients.

Theorem 4.9. (Classification via Laurent coefficients) *Let*

(1) *z_0 be an isolated singularity of f, and*
(2) *$f(z) = \displaystyle\sum_{n \in \mathbb{Z}} c_n (z - z_0)^n$ for $0 < |z - z_0| < R$, for some $R > 0$.*

Then

z_0 is removable	⇔	*For all $n < 0$, $c_n = 0$*
z_0 is a pole	⇔	*There exists an $m \in \mathbb{N}$ such that* *(a) $c_{-m} \neq 0$ and* *(b) for all $n < -m$, $c_n = 0$* *Then the order of the pole z_0 is m.*
z_0 is essential	⇔	*There are infinitely many* *negative indices n such that $c_n \neq 0$.*

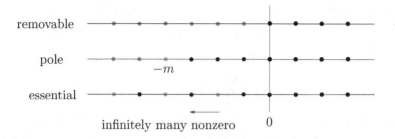

removable

pole

$-m$

essential

infinitely many nonzero   0

Proof. (1) $\boxed{z_0 \text{ is removable} \Rightarrow (\text{for all } n < 0, \, c_n = 0)}$.

Suppose that z_0 is a removable singularity. Then f has a holomorphic extension F defined for $|z - z_0| < R$. But then this holomorphic F has a Taylor series expansion

$$F(z) = \sum_{n=0}^{\infty} \widetilde{c}_n (z - z_0)^n \text{ for } |z - z_0| < R.$$

In particular, for $0 < |z - z_0| < R$, we obtain

$$f(z) = \sum_{n=0}^{\infty} \widetilde{c}_n (z - z_0)^n = \sum_{n \in \mathbb{Z}} c_n (z - z_0)^n,$$

and by the uniqueness of the Laurent series expansion in an annulus, it follows that $c_n = \widetilde{c}_n$ for $n \geq 0$, and $c_n = 0$ for all $n < 0$.

$\boxed{(\text{For all } n < 0, \, c_n = 0) \Rightarrow z_0 \text{ is removable}}$.

Suppose that $c_n = 0$ for all $n < 0$. Then for $0 < |z - z_0| < R$,

$$f(z) = \sum_{n \in \mathbb{Z}} \widetilde{c}_n (z - z_0)^n = \sum_{n=0}^{\infty} c_n (z - z_0)^n,$$

and so defining F for $|z - z_0| < R$ by

$$F(z) = \sum_{n=0}^{\infty} c_n (z - z_0)^n,$$

we see that F is holomorphic in $\{z \in \mathbb{C} : |z - z_0| < R\}$ (because it is a power series!) and moreover $F = f$ for $0 < |z - z_0| < R$.

(2) We will show that if z_0 is a pole of order m, then

$$f(z) = \frac{c_{-m}}{(z - z_0)^m} + \cdots + \frac{c_{-1}}{z - z_0} + c_0 + \sum_{n=0}^{\infty} c_n (z - z_0)^n,$$

with $c_{-m} \neq 0$. Suppose z_0 is a pole of order m. Using Theorem 4.8,

$$\lim_{z \to z_0} (z - z_0)\Big((z - z_0)^m f(z)\Big) = \lim_{z \to z_0} (z - z_0)^{m+1} f(z) = 0,$$

the function $(z - z_0)^m f(z)$ has a removable singularity at z_0. We have

$$(z - z_0)^m f(z) = (z - z_0)^m \sum_{n \in \mathbb{Z}} c_n (z - z_0)^n = \sum_{n \in \mathbb{Z}} c_n (z - z_0)^{n+m},$$

and by the previous part, this last series has all coefficients of negative powers of $z - z_0$ equal to 0, that is, $0 = c_{-(m+1)} = c_{-(m+2)} = \cdots$. Hence for $0 < |z - z_0| < R$,

$$(z - z_0)^m f(z) = c_{-m} + c_{-m+1}(z - z_0) + c_{-m+2}(z - z_0)^2 + \cdots \qquad (4.6)$$

and so

$$f(z) = \frac{c_{-m}}{(z - z_0)^m} + \cdots + \frac{c_{-1}}{(z - z_0)} + c_0 + c_1(z - z_0) + c_2(z - z_0)^2 + \cdots.$$

Moreover, the function $c_{-m} + c_{-m+1}(z - z_0) + c_{-m+2}(z - z_0)^2 + \cdots$ is holomorphic, and so it has the limit c_{-m} as $z \to z_0$. Thus

$$\lim_{z \to z_0} (z - z_0)^m f(z) = \lim_{z \to z_0} (c_{-m} + c_{-m+1}(z - z_0) + c_{-m+2}(z - z_0)^2 + \cdots) = c_{-m}.$$

On the other hand, since z_0 is a pole of order m, we know that

$$\neg \Big(\lim_{z \to z_0} (z - z_0)^m f(z) = 0 \Big).$$

Consequently, (4.6) implies that $c_{-m} \neq 0$.

Suppose now that

$$f(z) = \frac{c_{-m}}{(z - z_0)^m} + \cdots + \frac{c_{-1}}{z - z_0} + c_0 + \sum_{n=0}^{\infty} c_n (z - z_0)^n,$$

with $c_{-m} \neq 0$. We will show that z_0 pole of order m.

Suppose that there is some $m \in \mathbb{N}$ such that $c_{-m} \neq 0$ and $c_n = 0$ for all $n < -m$. Then $(z - z_0)^m f(z) = c_{-m} + c_{-m+1}(z - z_0) + c_{-m+2}(z - z_0)^2 + \cdots$, and since the right hand side defines a holomorphic function, say h, in $\{z \in \mathbb{C} : |z - z_0| < R\}$, it follows that

$$\lim_{z \to z_0} (z - z_0)^m f(z) = \lim_{z \to z_0} h(z) = h(z_0) = c_{-m} \neq 0, \quad \text{and}$$

$$\lim_{z \to z_0} (z - z_0)^{m+1} f(z) = 0 \cdot c_{-m} = 0.$$

Thus z_0 is a pole of order m of f, by Theorem 4.8.

(3) This is immediate from the previous two parts and the fact that an essential singularity is neither a removable singularity nor a pole. $\qquad \Box$

Let us reconsider Example 4.14.

Example 4.16.

(1) We have for $z \neq 0$,

$$\frac{\sin z}{z} = \frac{1}{z}\left(z - \frac{z^3}{3!} + \frac{z^5}{5!} - \frac{z^7}{7!} + - \cdots\right) = 1 - \frac{z^2}{3!} + \frac{z^4}{5!} - \frac{z^6}{7!} + - \cdots .$$

Since there are no negative powers of z appearing in the Laurent series expansion, it follows that 0 is a removable singularity of $\frac{\sin z}{z}$.

(2) For $z \neq 0$,

$$\frac{1}{z^3} = \cdots + 0 + \frac{1}{z^3} + 0 + \cdots .$$

Thus $c_{-3} = 1 \neq 0$, while $0 = c_{-4} = c_{-5} = \cdots$. So 0 is a pole of $\frac{1}{z^3}$ of order 3.

(3) For $z \neq 0$,

$$\exp\frac{1}{z} = 1 + \frac{1}{1!z} + \frac{1}{2!z^2} + \frac{1}{3!z^3} + \cdots .$$

For infinitely many negative indices n (in fact for all of them),

$$c_{-n} = \frac{1}{n!} \neq 0.$$

Thus 0 is an essential singularity of $\exp\frac{1}{z}$. $\lozenge$

Exercise 4.30. Let D be a domain and f be holomorphic in D such that f has only one zero z_0 in D, and the order of z_0 is $m > 1$. Show that the function $z \mapsto 1/f(z)$ has a pole of order m at z_0.

Exercise 4.31. Let D be a disc with center z_0. Suppose that f is nonzero and holomorphic in $D \setminus \{z_0\}$, and that f has a pole of order m at z_0. Show that the function $z \mapsto 1/f(z)$ has a holomorphic extension g to D, and that g has a zero of order m at z_0.

Exercise 4.32. Let D be a domain and $z_0 \in D$. Suppose that f has a pole of order m at z_0 and that f has the Laurent series expansion

$$f(z) = \sum_{n \in \mathbb{Z}} c_n(z - z_0)^n \quad \text{for } 0 < |z - z_0| < R,$$

where R is some positive number. Show that

$$c_{-1} = \frac{1}{(m-1)!} \lim_{z \to z_0} \frac{d^{m-1}}{dz^{m-1}}\left((z - z_0)^m f(z)\right).$$

Exercise 4.33. True or false?

(1) If f has a Laurent expansion $z^{-1}+c_0+c_1z+\cdots$, convergent in some punctured disc about the origin, then f has a pole at 0.

(2) A function may have different Laurent series centered at z_0, depending on the annulus of convergence selected.

(3) If f has an isolated singularity at z_0, then it has a Laurent series centered at z_0 and is convergent in some punctured disc $0 < |z - z_0| < R$.

(4) If a Laurent series for f convergent in some annulus $R_1 < |z - z_0| < R_2$ is actually a Taylor series (no negative powers of $z - z_0$), then this series actually converges in the full disc given by $|z - z_0| < R_2$ (at least).

(5) If the last conclusion holds, then f has at worst removable singularities in the full disc given by $|z - z_0| < R_2$ and may be considered holomorphic throughout this disc.

Exercise 4.34. Decide the nature of the singularity, if any, at 0 for the following functions. If the function is holomorphic or the singularity is isolated, expand the function in appropriate powers of z convergent in a punctured disc given by $0 < |z| < R$.

$$\sin z, \quad \sin \frac{1}{z}, \quad \frac{\sin z}{z}, \quad \frac{\sin z}{z^2}, \quad \frac{1}{\sin \frac{1}{z}}, \quad z \sin \frac{1}{z}$$

Exercise 4.35. True or false?

(1) $\displaystyle\lim_{z\to 0} \left| \exp \frac{1}{z} \right| = +\infty.$

(2) If f has a pole of order m at z_0, then there exists a polynomial p such that $f - \dfrac{p}{(z - z_0)^m}$ has a holomorphic extension to a disc around z_0.

(3) If f is holomorphic in a neighbourhood of 0, then there is an integer m such that $\dfrac{f}{z^n}$ has a pole at 0 whenever $n > m$.

(4) If f, g have poles of order m_f, m_g respectively at z_0, then their pointwise product $f \cdot g$ has a pole of order $m_f + m_g$ at z_0.

Exercise 4.36. Give an example of a function holomorphic in all of $\mathbb{C}$ except for essential singularities at the two points 0 and 1.

Exercise 4.37. The function f given by $f(z) = 1/(z - 1)$ clearly does not have a singularity at 0. As it has the Laurent series $z^{-1} + z^{-2} + z^{-3} + \cdots$ for $|z| > 1$, one might then say that this series has infinitely many negative powers of z, and fallaciously conclude that the point 0 is an essential singularity of f. Point out the flaw in this argument.

Exercise 4.38. Prove or disprove: If f and g have a pole and an essential singularity respectively at the point z_0, then fg has an essential singularity at z_0.

4.8.1 *Wild behaviour near essential singularities*

We now show a result which illustrates the "wild" behaviour of a function f at its essential singularity z_0. It says that given any complex number w, any $\epsilon > 0$, and any arbitrary small punctured disc Δ with center z_0, there is a point z in Δ such that $f(z)$ lies within a distance ϵ from w. So the image of any punctured disc centered at the essential singularity is dense in $\mathbb{C}$. Or in even more descriptive terms, f comes arbitrarily close to any complex value in every neighbourhood of z_0.

Theorem 4.10. (**"Casorati-Weierstrass"**[3]) *Suppose z_0 is an essential singularity of f. Then*

(1) *for every complex number w,*
(2) *for every $\delta > 0$, and*
(3) *for every $\epsilon > 0$,*

there exists a $z \in \mathbb{C}$ such that $|z - z_0| < \delta$ and $|f(z) - w| < \epsilon$.

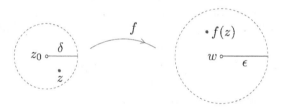

Proof. Suppose the statement is false. Then there exist $w \in \mathbb{C}$, $\epsilon > 0$ and $\delta > 0$ such that whenever $z \in D := \{z \in \mathbb{C} : 0 < |z - z_0| < \delta\}$, we have $|f(z) - w| \geq \epsilon$. Then g defined by

$$g(z) = \frac{1}{f(z) - w}$$

for $z \in D$ is holomorphic there, and $\lim_{z \to z_0} (z - z_0)g(z) = 0$, since

$$0 \leq |g(z)| = \frac{1}{|f(z) - w|} \leq \frac{1}{\epsilon}, \quad z \in D.$$

---

[3]This result was published by Weierstrass in 1876 (in German) and by the Sokhotski in 1873 (in Russian). So it was called Sokhotski's theorem in the Russian literature and Weierstrass's theorem in the Western literature. The same theorem was published by Casorati in 1868, and by Briot and Bouquet in the first edition of their book (1859), called *Theorie des fonctions doublement periodiques, et en particulier, des fonctions elliptiques*. However, Briot and Bouquet removed this theorem from the second edition (1875).

So g has a removable singularity at z_0. Let its extension be denoted again by g. Let m be the order of the zero of g at z_0. (Set $m = 0$ if $g(z_0) \neq 0$.) Then $g(z) = (z - z_0)^m h(z)$, for some function h holomorphic in D and such that $h(z_0) \neq 0$. Then for $0 < |z - z_0| < \delta$

$$
\begin{aligned}
(z - z_0)^{m+1} f(z) &= (z - z_0)^{m+1}(f(z) - w + w) \\
&= (z - z_0)^{m+1} \frac{1}{g(z)} + (z - z_0)^{m+1} \cdot w \\
&= \frac{z - z_0}{h(z)} + (z - z_0)^{m+1} \cdot w \\
&\xrightarrow{z \to z_0} \frac{0}{h(z_0)} + 0 \cdot w = 0.
\end{aligned}
$$

Thus either f has a removable singularity at z_0 (when $m = 0$) or a pole at z_0 (when $m \in \mathbb{N}$). Hence z_0 can't be an essential singularity of f, a contradiction. $\qquad \square$

Example 4.17. The function $\exp(1/z)$ has an essential singularity at 0. We show that it takes on any given nonzero w $(= \rho \exp(i\theta) \in \mathbb{C}, \, \rho, \theta \in \mathbb{R})$, in any arbitrarily small neighbourhood of 0. Setting $z = r \exp(it)$, we need to solve

$$
\exp \frac{1}{z} = \exp \left(\frac{\cos t}{r} - i \frac{\sin t}{r} \right) = \rho e^{i\theta}.
$$

By equating the absolute values, we obtain

$$
\frac{\cos t}{r} = \log \rho.
$$

On the other hand, by looking at arguments, we see that *a* solution is given when

$$
-\frac{\sin t}{r} = \theta.
$$

Using $(\cos t)^2 + (\sin t)^2 = 1$, we have

$$
r = \frac{1}{\sqrt{(\log \rho)^2 + \theta^2}}.
$$

Moreover

$$
\tan t = -\frac{\theta}{\log \rho}.
$$

But we are allowed to increase θ by integral multiplies of 2π, without changing w. Bearing this in mind, it is clear from the above expression for r that we can make r as small as we please. $\qquad \diamond$

The above example illustrates a much stronger theorem than the Casorati-Weierstrass Theorem, due to Picard, which says that the image of any punctured disc centered at an essential singularity misses at most one point of $\mathbb{C}$! In our example above, the exceptional value is $w = 0$. A proof of Picard's Theorem is beyond the scope of these notes, but can be found in the book [Conway (1978)].

Exercise 4.39. Prove, using the Casorati-Weierstrass Theorem, that if f has an essential singularity at z_0, and if w is any complex value whatever, then there exists a sequence $z_1, z_2, z_3, \cdots$ such that

$$\lim_{n \to \infty} z_n = z_0 \quad \text{and} \quad \lim_{n \to \infty} f(z_n) = w.$$

4.9 Residue Theorem

Suppose that D is a domain and that a holomorphic $f : D \setminus \{z_0\} \to \mathbb{C}$ has an isolated singularity at z_0. Let

$$f(z) = \sum_{n \in \mathbb{Z}} c_n (z - z_0)^n \text{ for } 0 < |z - z_0| < R.$$

Then we call the coefficient c_{-1} *the residue of f at z_0*, and denote it by

$$\text{res}(f, z_0).$$

Why the name? In everyday language, "residue" means something which is "left over". In the above too, one can think of c_{-1} as something which is left over in the following manner. We know that

$$2\pi i c_{-1} = \int_{C_r} \frac{f(z)}{(z - z_0)^{-1+1}} dz = \int_{C_r} f(z) dz = \int_{C_r} \left(\sum_{n \in \mathbb{Z}} c_n (z - z_0)^n \right) dz,$$

where C_r is given by $C_r(t) = z_0 + r \exp(it)$, $\in [0, 2\pi]$, and $r < R$. Note that we have

$$\int_{C_r} \left(c_n (z - z_0)^n \right) dz = \begin{cases} 0 & \text{for } n \neq -1, \\ 2\pi i c_{-1} & \text{for } n = -1, \end{cases}$$

and so if we imagine integrating termwise formally in the Laurent series expansion of f, then $2\pi i c_{-1}$ is what is left over.

What is the big fuss about this residue? The equation

$$2\pi i c_{-1} = \int_{C_r} f(z) dz$$

gives a way of computing contour integrals via calculating the residue of f at z_0 (which amounts to finding the value of the coefficient c_{-1} in the

Laurent expansion of f). So if there is way of calculating c_{-1} easily, then we can compute

$$\int_{\gamma} f(z)dz \ \left(= \int_{C_r} f(z)dz\right) \ = 2\pi i c_{-1} \tag{4.7}$$

for any closed path γ which is $D\setminus\{z_0\}$-homotopic to C_r. And we have seen in Exercise 4.32 that there *is* a way of calculating c_{-1} via the formula

$$c_{-1} = \frac{1}{(m-1)!} \lim_{z\to z_0} \frac{d^{m-1}}{dz^{m-1}}\left((z-z_0)^m f(z)\right),$$

when z_0 is a pole of order m of f. It turns out that some awkward real integrals can be computed by first relating them to a contour integral

$$\int_{\gamma} f(z)dz$$

for an appropriate holomorphic f and some path γ, and then using this route via the residue to evaluate the contour integral, and eventually also the awkward real integral. Here is an example.

Example 4.18. Consider the real integral

$$\int_0^{2\pi} \frac{1}{5+3\cos\theta}\,d\theta.$$

We view this as the contour integral along a circular path as follows. First we write

$$\cos\theta = \frac{\exp(i\theta)+\exp(-i\theta)}{2} = \frac{z+\dfrac{1}{z}}{2} = \frac{z^2+1}{2z},$$

where $z := \exp(i\theta)$. So if γ is the circular path with radius 1 and center at 0 defined by

$$\gamma(\theta) = \exp(i\theta), \quad \theta \in [0, 2\pi],$$

then $\gamma'(\theta)d\theta = i\exp(i\theta)d\theta = izd\theta$, and so

$$\int_0^{2\pi} \frac{1}{5+3\cos\theta}\,d\theta = \int_{\gamma} \frac{1}{5+3\cdot\left(\dfrac{z^2+1}{2z}\right)}\frac{1}{iz}\,dz = \int_{\gamma} -\frac{2i}{(3z+1)(z+3)}\,dz.$$

Let f be the function defined by

$$f(z) = -\frac{2i}{(3z+1)(z+3)}.$$

Then f has two poles, one at $-1/3$ and the other at -3, both of order 1. Of these, only the one at $-1/3$ lies inside γ. Thus,

$$\int_0^{2\pi} \frac{1}{5 + 3\cos\theta}d\theta = \int_\gamma f(z)dz = 2\pi i \cdot \operatorname{res}\left(f, -\frac{1}{3}\right).$$

What is c_{-1}? We have in a punctured disc with center $-1/3$ that

$$f(z) = \frac{c_{-1}}{z + \dfrac{1}{3}} + c_0 + c_1\left(z + \frac{1}{3}\right) + c_2\left(z + \frac{1}{3}\right)^2 + \cdots,$$

and so $\left(z + \dfrac{1}{3}\right)f(z) = c_{-1} + c_0\left(z + \dfrac{1}{3}\right) + \ldots$. Thus

$$c_{-1} = \lim_{z \to -1/3}\left(z + \frac{1}{3}\right)f(z) = \lim_{z \to -1/3} -\frac{2i}{3(z+3)} = -\frac{i}{4}.$$

Consequently, $\displaystyle\int_0^{2\pi} \frac{1}{5 + 3\cos\theta}d\theta = \frac{\pi}{2}.$ $\diamond$

More generally, if f has a finite number of *poles* in D, a result similar to (4.7) holds, and this is the content of the following.

Theorem 4.11. (Residue Theorem) *Let*

(1) *D be a domain,*
(2) *f be holomorphic in $D \setminus \{p_1, \ldots, p_K\}$,*
(3) *f have poles at $p_1, \ldots, p_K$ of order $m_1, \ldots, m_K$, respectively,*
(4) *γ be a closed path in $D \setminus \{p_1, \ldots, p_K\}$ and*
(5) *γ be such that for each $k = 1, \ldots, K$, γ is $D \setminus \{p_k\}$-homotopic to a circle C_k centered at p_k such that the interior of C_k is contained in D and contains only the pole p_k.*

Then $\displaystyle\int_\gamma f(z)dz = 2\pi i \sum_{k=1}^K \operatorname{res}(f, p_k).$

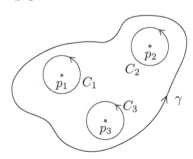

Proof. For each $k = 1, \ldots, K$, we can write

$$f(z) = \sum_{n=1}^{m_k} c_{-n,k}(z - p_k)^{-n} + \sum_{n=0}^{\infty} c_{n,k}(z - p_k)^n = f_k(z) + h_k(z),$$

$0 < |z - p_k| < R_k$, for some $R_k > 0$, where the sum with negative powers of $z - p_k$ is denoted by f_k, and the sum with nonnegative powers of $z - p_k$ is denoted by h_k. Note that h_k is holomorphic for $|z - p_k| < R_k$, and that f_k is a rational function, defined in $\mathbb{C} \setminus \{p_k\}$, having only one singularity, namely a pole at p_k. Thus $f - f_k$ is holomorphic in a small disc around p_k.

Set $g := f - (f_1 + \cdots + f_K)$. Fix a $k \in \{1, \ldots, K\}$. Write

$$g = (f - f_k) - \sum_{j \neq k} f_j.$$

Observe that both $f - f_k$ and each f_j for $j \neq k$ is holomorphic in a small disc around p_k. Thus g is holomorphic in a small disc around p_k. This happens with each $k \in \{1, \ldots, K\}$. Thus, g is holomorphic in D. We note that as γ is $D \setminus \{p_1\}$-homotopic to a circle C_1 centered at p_1, by the Cauchy Integral Theorem,

$$\int_{\gamma} g(z)dz = 0,$$

that is, $\displaystyle\int_{\gamma} \Big(f - (f_1 + \cdots + f_K) \Big) dz = 0$. By the Cauchy Integral Theorem,

$$\int_{\gamma} f_k(z)dz = \int_{C_k} f(z)dz = 2\pi i c_{-1,k}, \quad k = 1, \cdots, K.$$

So $\displaystyle\int_{\gamma} f(z)dz = \sum_{k=1}^{K} \int_{\gamma} f_k(z)dz = \sum_{k=1}^{K} 2\pi i c_{-1,k} = 2\pi i \sum_{k=1}^{K} \mathrm{res}(f, p_k).$ $\square$

Exercise 4.40. Evaluate $\displaystyle\int_{\gamma} \frac{\mathrm{Log}(z)}{1 + \exp z} dz$ along the path γ shown below.

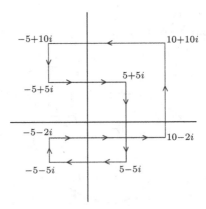

Integral of a real rational function using the Residue Theorem.
As we mentioned earlier, the Residue Theorem can be used to calculate
contour integrals, and sometimes gives an easy way to calculate some real
integrals. Let us see how it can be used to calculate the improper integrals
of rational functions. Consider a real integral of the type

$$\int_{-\infty}^{\infty} f(x)dx.$$

Such an integral, for which the interval of integration is not finite, is called
an *improper integral*, and it is defined as

$$\int_{-\infty}^{\infty} f(x)dx = \lim_{a \to -\infty} \int_{a}^{0} f(x)dx + \lim_{b \to +\infty} \int_{0}^{b} f(x)dx, \qquad (4.8)$$

when both the limits on the right hand side exist. In this case, also

$$\int_{-\infty}^{\infty} f(x)dx = \lim_{r \to +\infty} \int_{-r}^{r} f(x)dx. \qquad (4.9)$$

We call the right hand side in (4.9), if it exists, the *Cauchy principal value*
of the integral. (However, it can happen that the Cauchy principal value
exists, for example,

$$\lim_{r \to +\infty} \int_{-r}^{r} xdx = \lim_{r \to +\infty} \left(\frac{r^2}{2} - \frac{r^2}{2} \right) = 0,$$

but the improper integral doesn't exist: $\int_{0}^{b} xdx = \frac{b^2}{2}$.)

We assume that the function f in (4.8) is a real rational function whose
denominator is different from 0 for all real x, and the denominator has
degree at least two more than the degree of the numerator. Then it can
be seen that the limits on the right hand side in (4.8) exist, and so we can
start from (4.9).

Consider $\int_{\gamma} f(z)dz$, around a path γ shown in Figure 4.4.

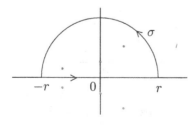

Fig. 4.4 γ consists of the semicircular arc σ and the straight line joining $-r$ to r.

Since f is rational, $z \mapsto f(z)$ has finitely many poles in the upper half plane, and if we choose r large enough, γ encloses all of these poles in its interior. By the Residue Theorem, we then obtain

$$\int_\gamma f(z)dz = \int_\sigma f(z)dz + \int_{-r}^r f(x)dx = 2\pi i \sum_{k:\ \mathrm{Im}(p_k)>0} \mathrm{res}(f, p_k),$$

where the sum consists of terms for all the poles that lie in the upper half-plane. From this, we obtain

$$\int_{-r}^r f(x)dx = 2\pi i \sum_{k:\ \mathrm{Im}(p_k)>0} \mathrm{res}(f, p_k) - \int_\sigma f(z)dz.$$

We show that as r increases, the value of the integral over the corresponding semicircular arc σ approaches 0. Indeed, from the fact that the degree of the denominator of f is at least two more than the degree of the numerator, it follows that there are M, r_0 large enough such that

$$|f(z)| < \frac{M}{|z|^2} \quad (|z| > r_0).$$

Hence for $r > r_0$,

$$\left| \int_\sigma f(z)dz \right| \leq \frac{M}{r^2}\pi r = \frac{M\pi}{r}.$$

Consequently, $\displaystyle\lim_{r \to +\infty} \int_\sigma f(z)dz = 0$, and so

$$\int_{-\infty}^\infty f(x)dx = 2\pi i \sum_{k:\ \mathrm{Im}(p_k)>0} \mathrm{res}(f, p_k).$$

Let us see an example of this method in action.

Example 4.19. We will show that $\displaystyle\int_0^\infty \frac{1}{1+x^4}dx = \frac{\pi}{2\sqrt{2}}$.

The function f given by $f(z) = \dfrac{1}{1+z^4}$ has four poles of order 1:

$$p_1 = \exp\left(\frac{\pi i}{4}\right), \quad p_2 = \exp\left(\frac{3\pi i}{4}\right), \quad p_3 = \exp\left(\frac{5\pi i}{4}\right), \quad p_4 = \exp\left(\frac{7\pi i}{4}\right).$$

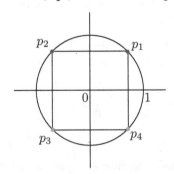

The first two of these poles lie in the upper half plane. We have

$$\operatorname{res}(f, p_1) = \lim_{z \to p_1} \frac{z - p_1}{1 + z^4} = \lim_{z \to p_1} \frac{1}{\dfrac{(1 + z^4) - (1 + p_1^4)}{z - p_1}}$$

$$= \frac{1}{\dfrac{d}{dz}(1 + z^4)\Big|_{z = p_1}} = \frac{1}{4p_1^3} = -\frac{1}{4}\exp\left(\frac{\pi i}{4}\right),$$

$$\operatorname{res}(f, p_2) = \lim_{z \to p_2} \frac{z - p_2}{1 + z^4} = \lim_{z \to p_2} \frac{1}{\dfrac{(1 + z^4) - (1 + p_2^4)}{z - p_2}}$$

$$= \frac{1}{\dfrac{d}{dz}(1 + z^4)\Big|_{z = p_2}} = \frac{1}{4p_2^3} = \frac{1}{4}\exp\left(-\frac{\pi i}{4}\right).$$

Thus

$$\int_{-\infty}^{\infty} \frac{1}{1 + x^4}\,dx = 2\pi i \left(-\frac{1}{4}\exp\left(\frac{\pi i}{4}\right) + \frac{1}{4}\exp\left(-\frac{\pi i}{4}\right)\right) = \pi \sin\frac{\pi}{4} = \frac{\pi}{\sqrt{2}}.$$

As f is even $(f(x) = f(-x)$ for all $x \in \mathbb{R})$, we have

$$\int_0^{\infty} \frac{1}{1 + x^4}\,dx = \frac{1}{2}\int_{-\infty}^{\infty} \frac{1}{1 + x^4}\,dx = \frac{\pi}{2\sqrt{2}}.$$

$\Diamond$

Here is an example of the computation of an exotic integral where the integrand is not a rational function.

Example 4.20. (Fresnel Integrals[4]) We will show that

$$\int_0^{\infty} \cos(x^2)\,dx = \int_0^{\infty} \sin(x^2)\,dx = \frac{\sqrt{\pi}}{2\sqrt{2}}.$$

We consider $\displaystyle\int_{\gamma} \exp(iz^2)\,dz$, where γ is shown below.

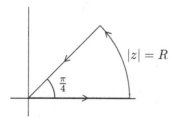

[4]These integrals arise in optics, in the description of diffraction phenomena.

Since $\exp(iz^2)$ is entire, by the Cauchy Integral Theorem we have

$$
\begin{aligned}
0 &= \int_\gamma \exp(iz^2)dz \\
&= \int_0^R \exp(ix^2)dx + \int_0^{\frac{\pi}{4}} \exp(iR^2\exp(2i\theta))iR\exp(i\theta)d\theta \\
&\quad - \int_0^R \exp\left(it^2\exp\left(i\frac{\pi}{2}\right)\right)\exp\left(i\frac{\pi}{4}\right)dt.
\end{aligned}
$$

We will show that the middle integral goes to 0 as $R \to \infty$. First note that

$$
\begin{aligned}
\left|\exp(iR^2\exp(2i\theta))iR\exp(i\theta)\right| &= \left|R\exp(R^2(i\cos(2\theta) - \sin(2\theta)))\right| \\
&= Re^{-R^2\sin(2\theta)}.
\end{aligned}
$$

But Figure 4.5 shows that whenever the angle t is such that $0 < t < \dfrac{\pi}{2}$,

$$
\frac{2}{\pi} \le \frac{\sin t}{t}.
$$

Indeed, the length of the arc PAQ, which is $2t$, is clearly less than the length of the semicircular arc PBQ, which is $\pi\sin t$, and so the inequality above follows. (Alternately, we could note that $t \mapsto \sin t$ is concave in $[0, \pi]$, because its second derivative is $-\sin t$, which is nonpositive in $[0, \pi]$, and so the graph of $\sin t$ is lies above that of the straight line graph of $2t/\pi$ joining the two points $(0,0)$ and $(\pi/2, 1)$.)

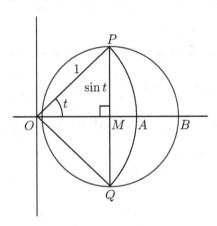

Fig. 4.5   Here P is any point on the circle with center 0 and radius 1 such that OP makes an angle t with the positive real axis. We reflect P in the real axis to get the point Q, and let M be the intersection of PQ with the real axis. With M as center and radius PM, we draw a circle, meeting the real axis on the right of A at the point B.

Applying this inequality with $t = 2\theta$ yields

$$\left| \exp(iR^2 \exp(2i\theta))iR\exp(i\theta) \right| = Re^{-R^2 \sin(2\theta)} \le Re^{-R^2 \frac{4\theta}{\pi}},$$

and so

$$\left| \int_0^{\frac{\pi}{4}} \exp(iR^2 \exp(2i\theta))iR\exp(i\theta)d\theta \right| \le R \int_0^{\frac{\pi}{4}} e^{-4R^2 \frac{\theta}{\pi}}d\theta = \frac{\pi}{4R}(1 - e^{-R^2}),$$

which tends to 0 as $R \to +\infty$. Hence we obtain

$$\lim_{R \to +\infty} \int_0^R \exp(ix^2)dx = \lim_{R \to +\infty} \int_0^R \exp\left(it^2 \exp\left(i\frac{\pi}{2} \right) \right) \exp\left(i\frac{\pi}{4} \right)dt$$

$$= \frac{1+i}{\sqrt{2}} \int_0^\infty e^{-t^2}dt = \frac{(1+i)\sqrt{\pi}}{2\sqrt{2}}.$$

Here we have used the known[5] fact that

$$\int_0^\infty e^{-x^2}dx = \frac{\sqrt{\pi}}{2}.$$

So by equating real and imaginary parts,

$$\int_0^\infty \cos(x^2)dx = \int_0^\infty \sin(x^2)dx = \frac{\sqrt{\pi}}{2\sqrt{2}}.$$

$\Diamond$

Exercise 4.41. Evaluate $\displaystyle\int_0^{2\pi} \frac{\cos\theta}{5 + 4\cos\theta}d\theta$.

Exercise 4.42. Evaluate the following integrals:

(1) $\displaystyle\int_0^\infty \frac{1}{1+x^2}dx$.

(2) $\displaystyle\int_0^\infty \frac{1}{(a^2 + x^2)(b^2 + x^2)}dx$, where $a > b > 0$.

(3) $\displaystyle\int_0^\infty \frac{1}{(1+x^2)^2}dx$.

(4) $\displaystyle\int_0^\infty \frac{1+x^2}{1+x^4}dx$.

Exercise 4.43. If $n \in \mathbb{N}$ and C is the path $C(t) = \exp(i\theta)$, $\theta \in [0, 2\pi]$, find $\displaystyle\int_C \frac{\exp z}{z^{n+1}}dz$. Deduce that $\displaystyle\int_0^{2\pi} e^{\cos\theta}\cos(n\theta - \sin\theta)d\theta = \frac{2\pi}{n!}$.

---

[5]With $I := \displaystyle\int_0^\infty e^{-x^2}dx$, $I^2 = \left(\displaystyle\int_0^\infty e^{-x^2}dx \right)\left(\displaystyle\int_0^\infty e^{-y^2}dy \right) = \displaystyle\int_0^\infty \int_0^\infty e^{-(x^2+y^2)}dxdy$

$$= \int_0^{\frac{\pi}{2}} \int_0^\infty e^{-r^2}rdrd\theta = \frac{\pi}{4}.$$

Exercise 4.44. Let f have a zero of order 1 at z_0, so that $1/f$ has a pole of order 1 at z_0. Prove that $\operatorname{res}\left(\dfrac{1}{f}, z_0\right) = \dfrac{1}{f'(z_0)}$.

Exercise 4.45. Prove that $\operatorname{res}\left(\dfrac{1}{\sin z}, k\pi\right) = (-1)^k$.

Exercise 4.46. The nth Fibonacci number f_n, where $n \geq 0$, is defined by the following recurrence relation: $f_0 = 1$, $f_1 = 1$, $f_n = f_{n-1} + f_{n-2}$ for $n \geq 2$. Let $F(z) := \displaystyle\sum_{n=0}^{\infty} f_n z^n$.

(1) Prove by induction that $f_n \leq 2^n$ for all $n \in \mathbb{N}$.

(2) Using the estimate $f_n \leq 2^n$, deduce that the radius of convergence of F is at least $1/2$.

(3) Show that the recurrence among the f_n implies that $F(z) = \dfrac{1}{1 - z - z^2}$.

   Hint: Write down the Taylor series for $zF(z)$ and $z^2F(z)$ and add.

(4) Verify that $\operatorname{res}\left(\dfrac{1}{z^{n+1}(1 - z - z^2)}, 0\right) = f_n$.

(5) Using the Residue Theorem, prove

$$f_n = \frac{1}{\sqrt{5}}\left(\left(\frac{1 + \sqrt{5}}{2}\right)^{n+1} - \left(\frac{1 - \sqrt{5}}{2}\right)^{n+1}\right).$$

   Hint: Integrate $\dfrac{1}{z^{n+1}(1 - z - z^2)}$ around a circle with center 0 and radius R and show that this integral vanishes as $R \to +\infty$.

4.10   Notes

§4.8 follows closely [Beck, Marchesi, Pixton, Sabalka (2008)]. Exercise 4.11, 4.24, 4.33, 4.35, 4.36, 4.37, 4.38, 4.39 are taken from [Flanigan (1972)]. Exercise 4.14 is taken from [Volkovyskiĭ, Lunts, Aramanovich (1991)]. Exercise 4.15, 4.25 are taken from [Rudin (1987)]. Exercise 4.28 is taken from [Beck, Marchesi, Pixton, Sabalka (2008)]. Exercise 4.40 is taken from [Ash and Novinger (2007)].

Chapter 5

Harmonic functions

In this last chapter, we study:

(1) harmonic functions, which are real-valued functions that solve a certain partial differential equation, called the Laplace equation,

(2) that real and imaginary parts of holomorphic functions are harmonic, and that the converse holds locally and globally on simply connected domains,

(3) some consequences of the above interplay between harmonic and holomorphic functions, in particular in a certain "boundary value problem", called the Dirichlet problem.

5.1 What is a harmonic function?

Definition 5.1. Let U be an open subset of $\mathbb{R}^2$. A function $u : U \to \mathbb{R}$ is called *harmonic* if u has continuous partial derivatives of order 2 (abbreviated by writing $u \in C^2$), and it satisfies the *Laplace equation*:

$$(\Delta u)(x, y) := \frac{\partial^2 u}{\partial x^2}(x, y) + \frac{\partial^2 u}{\partial y^2}(x, y) = 0 \text{ for all } (x, y) \in U.$$

Example 5.1. Let $U = \mathbb{R}^2$. Consider the function $u : U \to \mathbb{R}$ given by $u(x, y) = x^2 - y^2$ for $(x, y) \in \mathbb{R}^2$. Then

$$\frac{\partial u}{\partial x} = 2x \quad \frac{\partial^2 u}{\partial x^2} = 2,$$

$$\frac{\partial u}{\partial y} = -2y \quad \frac{\partial^2 u}{\partial y^2} = -2.$$

Thus $\dfrac{\partial^2 u}{\partial x^2}(x, y) + \dfrac{\partial^2 u}{\partial y^2}(x, y) = 2 - 2 = 0$. Since $u \in C^2$ and $\Delta u = 0$ in $\mathbb{R}^2$, u is harmonic in $\mathbb{R}^2$. $\Diamond$

Of course, not all functions are harmonic.

Example 5.2. Consider the function $\widetilde{u}$ given by $\widetilde{u}(x, y) = x^2 + y^2$ for $(x, y) \in \mathbb{R}^2$. Then for all $(x, y) \in \mathbb{R}^2$,

$$\frac{\partial^2 \widetilde{u}}{\partial x^2}(x, y) + \frac{\partial^2 \widetilde{u}}{\partial y^2}(x, y) = 2 + 2 = 4 \neq 0.$$

Since $\Delta \widetilde{u}$ is never 0 in $\mathbb{R}^2$, $\widetilde{u}$ is not harmonic in any open subset of $\mathbb{R}^2$. ◊

Exercise 5.1. Show that the following functions u are harmonic in the corresponding open set U.

(1) $u(x, y) = \log(x^2 + y^2)$, $U = \mathbb{R}^2 \setminus \{(0, 0)\}$.

(2) $u(x, y) = e^x \sin y$, $U = \mathbb{R}^2$.

Exercise 5.2. Show that the set $\mathrm{Har}(U)$ of all harmonic functions on an open set U forms a real vector space with pointwise operations.

Exercise 5.3. Is the pointwise product of two harmonic functions also necessarily harmonic?

Why bother about harmonic functions?   Harmonic functions are important because they satisfy the Laplace equation, which is important among other things for two primary reasons:

(1) The Laplace equation is the prototype of an important class of PDEs, namely "elliptic equations", which is one of the three main classes of PDEs.

Class of PDE	Main example	
Elliptic	Laplace equation	$\dfrac{\partial^2 u}{\partial x^2} + \dfrac{\partial^2 u}{\partial y^2} = 0$
Parabolic	Diffusion equation	$\dfrac{\partial^2 u}{\partial t} - \dfrac{\partial^2 u}{\partial x^2} = 0$
Hyperbolic	Wave equation	$\dfrac{\partial^2 u}{\partial t^2} - \dfrac{\partial^2 u}{\partial x^2} = 0$

(2) The Laplace equation arises in many applications, for example in physics in the following scenarios. In hydrodynamics, the "velocity potential" of the fluid flow satisfies the Laplace equation, while in electrostatics, the electrostatic potential satisfies the Laplace equation. The Laplace equation also has an important link with stochastic processes. We describe this very roughly below.

Imagine the open unit disc $\mathbb{D} := \{z \in \mathbb{C} : |z| < 1\}$. Consider a particle starting at a point $z \in D$ and undergoing Brownian motion (for example think of a pollen grain in water, bombarded by many tiny water molecules producing "random" motion). Intuitively one feels that since the motion is random, eventually the particle will leave the boundary $\mathbb{T} := \{z \in \mathbb{C} : |z| = 1\}$ of the disc $\mathbb{D}$. Let us denote by ζ_z the point on $\mathbb{T}$ where the particle first exits the unit circle $\mathbb{T}$, having started at z. So we get a random variable ζ_z which lives on the unit circle. See Figure 5.1.

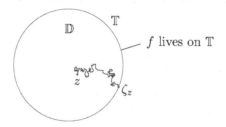

Fig. 5.1   Brownian motion and the Dirichlet problem.

Now let $f : \mathbb{T} \to \mathbb{R}$ be a given continuous function. Then we can think of $f(\zeta_z)$ as being a real-valued random variable on $\mathbb{T}$. Let us denote its expectation by $\mathbb{E}(f(\zeta_z))$. This depends on where one starts initially, that is, it depends on z. Let $u : \mathbb{D} \to \mathbb{R}$ be given by $u(z) = \mathbb{E}(f(\zeta_z))$, $z \in \mathbb{D}$. It turns out that u is then harmonic, and in fact it is a solution to the "Dirichlet problem", which is the boundary value problem, where given $f : \mathbb{T} \to \mathbb{R}$ on the boundary $\mathbb{T}$, we have to find a function u, solving the Laplace equation in the interior $\mathbb{D}$ of $\mathbb{T}$, such that it has a continuous extension to $\mathbb{T}$, matching with the given data f:

$$\begin{cases} \Delta u = 0 \text{ in } \mathbb{D}, \\ u|_{\mathbb{T}} = f. \end{cases}$$

5.2   What is the link between harmonic functions and holomorphic functions?

It might appear that harmonic functions should belong just to the realm of Real Analysis. In this section, we will now learn about two results, which will amply justify their study in Complex Analysis. Roughly a function is harmonic in an open set if and only if locally, it is the real part of some holomorphic function.

Theorem 5.1. *Let*

(1) *U be an open subset of $\mathbb{C}$ and*
(2) *$f : U \to \mathbb{C}$ be holomorphic in U.*

Then $\left.\begin{array}{l} u := \mathrm{Re}(f), \\ v := \mathrm{Im}(f) \end{array}\right\}$ *are harmonic functions in U.*

Vice versa, we will also learn the following converse to this.

Theorem 5.2. *Let*

(1) *U be simply connected and*
(2) *$u : U \to \mathbb{R}$ be harmonic in U.*

Then there exists a function $v : U \to \mathbb{R}$, such that v is harmonic in U and $f := u + iv$ is holomorphic in U.

Note that for the f in the conclusion of the above result, we have $\mathrm{Re}(f) = u$, and $\mathrm{Im}(f) = v$. So what we are saying is that every harmonic function in a simply connected domain is the real part of some holomorphic function defined there. Since a unit disc is simply connected, in particular it follows that every harmonic function is *locally* the real part of a holomorphic function defined (at least in that disc). We will see later on in Exercise 5.5 that the assumption of simply connectedness is not superfluous, and that given a harmonic function in a *non*-simply connected domain, there may fail to exist a *globally* defined holomorphic function in the whole domain whose real part is the given harmonic function. But we are jumping ahead. Let us first focus on the first result mentioned above. Before proving Theorem 5.1, let us revisit Example 5.1.

Example 5.3. In our previous example, when $U = \mathbb{R}^2$ and $u = x^2 - y^2$, we have $u = \mathrm{Re}(z^2) = \mathrm{Re}(x^2 - y^2 + 2xyi)$ and z^2 is entire. So Theorem 5.1 delivers again our old observation that u is harmonic in $\mathbb{R}^2$. In fact from our calculation, Theorem 5.1 also gives us that $v := 2xy = \mathrm{Im}(z^2)$ is harmonic. (We can of course check this by brute force:

$$\frac{\partial u}{\partial x} = 2y, \quad \frac{\partial^2 u}{\partial x^2} = 0, \quad \frac{\partial v}{\partial x} = 2x, \quad \frac{\partial^2 v}{\partial y^2} = 0,$$

and so $\dfrac{\partial^2 u}{\partial x^2} + \dfrac{\partial^2 v}{\partial y^2} = 0 + 0 = 0$.)

On the other hand, it follows from Example 5.2 and Theorem 5.1 that for any open subset U of $\mathbb{C}$, $\widetilde{u} := x^2 + y^2$ is not the real part of any holomorphic function defined in U. $\lozenge$

Proof. (of Theorem 5.1.) We have $f(x + iy) = u(x,y) + iv(x,y)$ for $(x,y) \in U$. Since f is infinitely many times differentiable, we know that u, v have partial derivatives of all orders, and so by the Cauchy-Riemann equations, we have

$$\frac{\partial^2 u}{\partial x^2} = \frac{\partial}{\partial x}\left(\frac{\partial u}{\partial x}\right) \overset{\text{(C-R)}}{=} \frac{\partial}{\partial x}\left(\frac{\partial v}{\partial y}\right) \overset{(u \in C^2)}{=} \frac{\partial}{\partial y}\left(\frac{\partial v}{\partial x}\right) \overset{\text{(C-R)}}{=} \frac{\partial}{\partial y}\left(-\frac{\partial u}{\partial y}\right)$$

$$= -\frac{\partial^2 u}{\partial y^2},$$

and so u is harmonic. Similarly,

$$\frac{\partial^2 v}{\partial x^2} = \frac{\partial}{\partial x}\left(\frac{\partial v}{\partial x}\right) \overset{\text{(C-R)}}{=} \frac{\partial}{\partial x}\left(-\frac{\partial u}{\partial y}\right) \overset{(v \in C^2)}{=} \frac{\partial}{\partial y}\left(-\frac{\partial u}{\partial x}\right) \overset{\text{(C-R)}}{=} \frac{\partial}{\partial y}\left(-\frac{\partial v}{\partial y}\right)$$

$$= -\frac{\partial^2 v}{\partial y^2},$$

and so v is harmonic as well. (Alternately, we could have also noted that $v = \text{Re}(-if)$.) $\square$

Now we show Theorem 5.2, which gives a converse to the above result when the open set in question is a simply connected domain. As mentioned earlier, for more general domains, it can happen that there are harmonic functions which aren't globally the real part of a holomorphic function; see Exercise 5.5, where we take

(1) $U := \mathbb{R}^2 \setminus \{(0,0)\}$ (which is not simply connected) and
(2) $u := \log(x^2 + y^2)$ (which is harmonic in U).

Then there is no holomorphic f in $\mathbb{C} \setminus \{0\}$ such that $u = \text{Re}(f)$ in U.

Proof. (of Theorem 5.2.) We will construct a holomorphic f with real part u, and then $v := \text{Im}(f)$ will serve as the required harmonic function. First set

$$g = \frac{\partial u}{\partial x} - i\frac{\partial u}{\partial y}.$$

We will prove that g is holomorphic, and then to construct a primitive of g, necessarily holomorphic, which will be the f we seek. To show that g

is holomorphic, we will use Theorem 2.2. First we note that since u is harmonic, the functions

$$\text{Re}(g) = \frac{\partial u}{\partial x} \quad \text{and} \quad \text{Im}(g) = -\frac{\partial u}{\partial y}$$

have continuous partial derivatives. Moreover, using the fact that u satisfies the Laplace equation, we can now see that the real and imaginary parts of g satisfy the Cauchy-Riemann equations:

$$\frac{\partial}{\partial x}(\text{Re}(g)) = \frac{\partial}{\partial x}\left(\frac{\partial u}{\partial x}\right) = \frac{\partial^2 u}{\partial x^2} = -\frac{\partial^2 u}{\partial y^2} = \frac{\partial}{\partial y}\left(-\frac{\partial u}{\partial y}\right) = \frac{\partial}{\partial y}(\text{Im}(g)),$$

$$\frac{\partial}{\partial y}(\text{Re}(g)) = \frac{\partial}{\partial y}\left(\frac{\partial u}{\partial x}\right) = \frac{\partial}{\partial x}\left(\frac{\partial u}{\partial y}\right) = -\frac{\partial}{\partial x}\left(-\frac{\partial u}{\partial y}\right) = -\frac{\partial}{\partial x}(\text{Im}(g)).$$

Hence g is holomorphic in U, and by Theorem 3.5, since U is simply connected, it has a primitive G in U. Decompose $G = \widetilde{u} + i\widetilde{v}$ into its real and imaginary parts $\widetilde{u}, \widetilde{v}$. Then

$$\frac{\partial u}{\partial x} - i\frac{\partial u}{\partial y} = g = G' = \frac{\partial \widetilde{u}}{\partial x} + i\frac{\partial \widetilde{v}}{\partial x}, \tag{5.1}$$

and so

$$\frac{\partial(u - \widetilde{u})}{\partial x} = 0,$$

and so it follows from the Fundamental Theorem of Integral Calculus, that $u - \widetilde{u}$ is locally constant along horizontal lines. Also, (5.1) gives

$$-\frac{\partial u}{\partial y} = \frac{\partial \widetilde{v}}{\partial x} \overset{\text{(C-R)}}{=} -\frac{\partial \widetilde{u}}{\partial y},$$

where have used the Cauchy-Riemann equations to obtain the last equality. Hence

$$\frac{\partial(u - \widetilde{u})}{\partial y} = 0,$$

showing that $u - \widetilde{u}$ is locally constant along vertical lines as well. Since U is a domain, any two points in U can be joined by a stepwise path, and so it follows from here that $\widetilde{u} - u$ must be a constant, say C ($\in \mathbb{R}$), in U. Consequently,

$$f := G - C = (\widetilde{u} - C) + i\widetilde{v} = u + i\widetilde{v},$$

and so we see that f is holomorphic in U, $u = \text{Re}(f)$, $v := \widetilde{v} = \text{Im}(G)$ is harmonic in U and $f = u + iv$. $\square$

Definition 5.2. Let U be an open set and $u : U \to \mathbb{R}$ be harmonic in U. Any harmonic function $v : U \to \mathbb{R}$ such that $f := u + iv$ is holomorphic in U is called a *harmonic conjugate* of u.

Example 5.4. $v := 2xy$ is a harmonic conjugate of the harmonic function $u := x^2 - y^2$ considered in Example 5.1 because

$$f := u + iv = x^2 - y^2 + 2xyi = (x + iy)^2 = z^2,$$

where $z = x + iy$, is entire. $\Diamond$

Harmonic conjugates are obviously not unique since we can just add a constant to a harmonic conjugate and get a new harmonic conjugate. In a simply connected domain, for each harmonic function, Theorem 5.2 guarantees the existence of a harmonic conjugate, but it is not particularly useful for finding a harmonic conjugate (from the above proof, we see that it relies on the construction of the primitive G). A more direct way is to look at the Cauchy-Riemann equations, as shown in the following example.

Example 5.5. Let $u = -(\sin x)(\sinh y)$. Then

$$\frac{\partial^2 u}{\partial x^2} = +(\sin x)(\sinh y),$$

$$\frac{\partial^2 u}{\partial y^2} = -(\sin x)(\sinh y),$$

and so u is harmonic in $\mathbb{R}^2$. As $\mathbb{R}^2$ is simply connected, we know that there is a harmonic conjugate. Can we find one? If v is a harmonic conjugate, then we know that $u + iv$ will be holomorphic, and so the pair u, v must satisfy the Cauchy-Riemann equations. So we need a v such that

$$\frac{\partial v}{\partial x} = -\frac{\partial u}{\partial y} = +(\sin x)(\cosh y), \tag{5.2}$$

$$\frac{\partial v}{\partial y} = \frac{\partial u}{\partial x} = -(\cos x)(\sinh y). \tag{5.3}$$

Integrating (5.2) with respect to x while keeping y fixed, we see that

$$v = -(\cos x)(\cosh y) + C(y), \tag{5.4}$$

for some constant $C(y)$ depending on y. So assuming for the moment that $C(y)$ is differentiable, we have

$$-(\cos x)(\sinh y) \overset{(5.3)}{=} \frac{\partial v}{\partial y} \overset{(5.4)}{=} -(\cos x)(\sinh y) + \frac{dC}{dy}(y).$$

Consequently

$$\frac{dC}{dy}(y) = 0,$$

and so $C(y) = C$ for all y. So we could take any constant C, and in particular we may choose $C = 0$. So based on the above rough reasoning, we take

$$v = -(\cos x)(\cosh y)$$

as a candidate for a harmonic conjugate of u. In order to verify this guess, we can quickly note that the Cauchy-Riemann equations *are* satisfied by the pair (u, v), so that $f := u + iv$ is holomorphic. But instead, let us use an old calculation we performed in Exercise 1.37 to check directly that $f := u + iv$ is holomorphic by actually finding the f:

$$f = u + iv = -(\sin x)(\sinh y) - i(\cos x)(\cosh y)$$
$$= -i\Big((\cos x)(\cosh y) - i(\sin x)(\sinh y)\Big) = -i\cos z,$$

which is entire.   ◊

Exercise 5.4. Find harmonic conjugates for the following harmonic functions in $\mathbb{R}^2$:

$$e^x \sin y, \quad x^3 - 3xy^2 - 2y, \quad x(1 + 2y).$$

Exercise 5.5. Show that there is no holomorphic function f defined in $\mathbb{C} \setminus \{0\}$ whose real part is the harmonic function u defined by $u(x, y) = \log(x^2 + y^2)$, $(x, y) \in \mathbb{R}^2 \setminus \{(0, 0)\}$.
Hint: If v is a harmonic conjugate of u, then also $h(z) := z^2 \exp(-(u + iv))$ is holomorphic. Find $|h|$, and conclude that $h' = 0$. Show that $h' = 0$ implies that $2/z$ has a primitive in $\mathbb{C} \setminus \{0\}$, which is impossible.

Exercise 5.6. Is it possible to find a $v : \mathbb{R}^2 \to \mathbb{R}$ so that f defined by

$$f(x + iy) = x^3 + y^3 + iv(x, y), \quad (x, y) \in \mathbb{R}^2,$$

is holomorphic in $\mathbb{C}$?

5.3   Consequences of the two way traffic: holomorphic ↔ harmonic

In the previous section, the two results given in Theorems 5.1 and 5.2 show that there is a two way traffic between the real analysis world of harmonic functions and the complex analysis world of holomorphic functions, allowing a fruitful interaction between the two worlds.

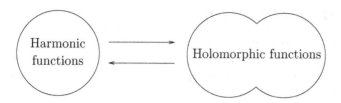

In the previous three chapters we have learnt many pleasant properties possessed by holomorphic functions. Let us now use some of these to derive some important properties of harmonic functions. In particular, we will show the following results in this section:

(1) If u is harmonic, then it is C^∞.
(2) The Mean Value Property for harmonic functions.
(3) The Maximum Principle for harmonic functions.
(4) Uniqueness of solutions to the Dirichlet Problem.

5.3.1 *Harmonic functions are smooth*

Corollary 5.1. *Harmonic functions are infinitely many times differentiable.*

(Note that the definition of a harmonic function demands only *twice* continuous differentiability. The remarkable result here says that thanks to the fact that the Laplace equation is satisfied, in fact the function has got to be infinitely differentiable. A result of this type is called a *regularity result* in PDE theory.)

Proof. Suppose that u is a harmonic function in an open set U. Let $z_0 = (x_0, y_0) \in U$. Then there is a $r > 0$ such that the disc D with center z_0 and radius r is contained in U. But $u|_D$ is harmonic in D and D is simply connected. So there is a holomorphic function f defined in D, such that $\mathrm{Re}(f) = u$ in D. But f is infinitely many times complex differentiable in D. Consequently, u is infinitely many times differentiable in D, and in particular at $z_0 \in D$. As the choice of $z_0 \in D$ was arbitrary, the result follows. $\qquad\square$

Exercise 5.7. Show that all partial derivatives of a harmonic function are harmonic.

5.3.2 Mean value property

Using the Cauchy Integral Formula, we immediately obtain the following "mean value property" of harmonic functions, which says that the value of a harmonic function is the average (or mean) of the values on a circle with that point as the center.

Theorem 5.3. (Mean-value property of harmonic functions) *Let*

(1) *U be an open set,*
(2) *$u : U \to \mathbb{R}$ be harmonic in U,*
(3) *$z_0 \in U$,*
(4) *$R > 0$ be such that the disc $\{z \in \mathbb{C} : |z - z_0| < R\} \subset U$.*

Then $u(z_0) = \dfrac{1}{2\pi} \displaystyle\int_0^{2\pi} u(z_0 + r\exp(it))dt$ for all r such that $0 < r < R$.

Proof. The disc $D := \{z \in \mathbb{C} : |z - z_0| < R\}$ is simply connected, and so there is a holomorphic function f defined in D, whose real part is u. But now by the Cauchy Integral Formula, if C is the circular path given by $C(t) = z_0 + r\exp(it)$ ($t \in [0, 2\pi]$), then

$$
\begin{aligned}
f(z_0) &= \frac{1}{2\pi i} \int_C \frac{f(z)}{z - z_0} dz = \frac{1}{2\pi i} \int_0^{2\pi} \frac{f(z_0 + r\exp(it))}{r\exp(it)} ir\exp(it)dt \\
&= \frac{1}{2\pi} \int_0^{2\pi} f(z_0 + r\exp(it))dt.
\end{aligned}
$$

Equating real parts, the claim is proved. $\square$

5.3.3 Maximum Principle

From the Maximum Modulus Theorem (see page 129), we also obtain the following.

Theorem 5.4. (Maximum Principle) *Suppose that*

(1) *U is a simply connected domain,*
(2) *$u : U \to \mathbb{R}$ is harmonic in U,*
(3) *$z_0 \in U$ is a point such that $u(z_0) \geq u(z)$ for all $z \in D$.*

Then u is constant in U.

Proof. There is a holomorphic function f defined in U whose real part equals u. But then the function $g : U \to \mathbb{C}$ defined by $g(z) = \exp(f(z))$

$(z \in U)$ is holomorphic too. We have

$$|g(z_0)| = |\exp(f(z_0))| = e^{\operatorname{Re}(z_0)} = e^{u(z_0)} \geq e^{u(z)} = |g(z)| \text{ for all } z \in U.$$

By the Maximum Modulus Theorem applied to g, it follows that g must be constant in U. Thus $|g|$ is also constant in U, that is, $|g| = e^{\operatorname{Re}(f)} = e^u$ is a constant in U. Taking the (real) logarithm, it follows that u is constant in U. $\qquad \square$

5.4 Uniqueness of solution for the Dirichlet problem

The Maximum Principle has an important consequence about the uniqueness of solutions to the Dirichlet problem, as explained below. Let

$$\mathbb{D} := \{z \in \mathbb{C} : |z| < 1\},$$
$$\mathbb{T} := \{z \in \mathbb{C} : |z| = 1\}.$$

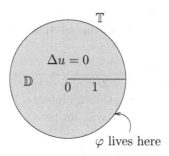

φ lives here

Then the *Dirichlet problem* is the following:

Given $\varphi : \mathbb{T} \to \mathbb{R}$, continuous,
find a $u : \mathbb{D} \cup \mathbb{T} \to \mathbb{R}$ such that
   (1) u is continuous on $\mathbb{D} \cup \mathbb{T}$,
   (2) $u|_\mathbb{T} = \varphi$,
   (3) u has continuous partial derivatives of order 2 in $\mathbb{D}$,
   (4) $\Delta u = 0$ in $\mathbb{D}$.

The given function φ is called the *boundary data*. The reason one is interested in solving the Dirichlet problem is that the need arises in applications, for example in heat conduction, electrostatics, and fluid flow. Using the Maximum Principle, we can show the following.

Proposition 5.1. *The solution to the Dirichlet problem is unique.*

Proof. Indeed, let u_1, u_2 be two distinct solutions corresponding to the boundary data φ. In particular, $u_1 = u_2$ on $\mathbb{T}$. So u_1 must differ from u_2 somewhere inside $\mathbb{D}$. Without loss of generality suppose that there is a point $w \in \mathbb{D}$ where $u_1(w) > u_2(w)$. (Otherwise exchange their labels.) Then $u := u_1 - u_2$ is such that

(1) $u = 0$ on $\mathbb{T}$,

(2) $u(w) > 0$,

(3) u is harmonic in $\mathbb{D}$.

Let $z_0 \in \mathbb{D} \cup \mathbb{T}$ be the maximizer for the real-valued continuous function u on the compact set $\mathbb{D} \cup \mathbb{T}$. From (1),(2), $z_0 \notin \mathbb{T}$. So $z_0 \in \mathbb{D}$. We have $u(z_0) \geq u(z)$ for all $z \in \mathbb{D}$, and so by the Maximum Principle, u must be constant in $\mathbb{D}$. But u is continuous on $\mathbb{D} \cup \mathbb{T}$, and u is 0 on $\mathbb{T}$. So it follows that the constant value of u must be 0 everywhere in $\mathbb{D} \cup \mathbb{T}$. Hence $u_1 = u_2$, a contradiction. $\square$

Remark 5.1. It can be shown that the following expression, called the *Poisson Integral Formula*, gives the solution to the Dirichlet problem with boundary data φ:

$$u(r \exp(it)) = \frac{1}{2\pi} \int_0^{2\pi} \frac{1 - r^2}{1 - 2r\cos(\theta - t) + r^2} \varphi(\exp(i\theta))d\theta \quad (\zeta \in \mathbb{T}).$$

This can be derived using the Cauchy Integral Formula, but there are some technical subtleties, and so we will not prove this here.

Exercise 5.8. (Some half-plane Dirichlet problems.) Given the "boundary data" $b : \mathbb{R} \to \mathbb{R}$, we consider the problem of finding a continuous, real-valued function h defined in the closed upper half-plane $y \geq 0$, such that h is harmonic in the open upper half-plane $y > 0$ and moreover, $h(x, 0) = b(x)$.

(1) If b is just a polynomial p, then show that we can simply take h given by $h(x, y) = \text{Re}(p(x + iy))$.

(2) Prove that if

$$b(x) = \frac{1}{1 + x^2},$$

then $(x, y) \mapsto \text{Re}(b(x + iy))$ is *not* a solution (because of the pole at $z = i$). Show that

$$h(x, y) := \text{Re}\left(\frac{i}{z + i}\right) = \frac{y + 1}{x^2 + (y + 1)^2}$$

gives a solution to the Dirichlet problem.

Exercise 5.9. Let $u : \mathbb{R}^2 \to \mathbb{R}$ be a harmonic function such that $u(x, y) > 0$ for all $(x, y) \in \mathbb{R}^2$. Prove that u is constant.

Hint: Let f be an entire function whose real part is u. Consider $\exp(-f)$.

Exercise 5.10. The regularity of functions satisfying the Laplace equation is not completely for free. Here is an example to show that a *discontinuous* function may satisfy the Laplace equation! Consider the function $u : \mathbb{R}^2 \to \mathbb{R}$ defined to be the real part of e^{-1/z^4} when $z \neq 0$ and 0 at the origin.

(1) Verify that u is discontinuous at 0.

(2) Check that $u(x, 0) = e^{-1/x^4}$, $u(0, y) = e^{-1/y^4}$.

(3) Being the real part of a holomorphic function in $\mathbb{C} \setminus \{0\}$, we know already that u satisfies the Laplace equation everywhere in $\mathbb{R}^2 \setminus \{(0, 0)\}$. Show that also

$$\frac{\partial^2 u}{\partial x^2}(0, 0) \quad \text{and} \quad \frac{\partial^2 u}{\partial y^2}(0, 0)$$

exist, and that $\dfrac{\partial^2 u}{\partial x^2}(0, 0) + \dfrac{\partial^2 u}{\partial y^2}(0, 0) = 0$.

Exercise 5.11. Let D_1, D_2 be domains in $\mathbb{C}$. Let $\varphi : D_1 \to D_2$ be holomorphic. Show that if $h : D_2 \to \mathbb{R}$ is harmonic, then $h \circ \varphi : D_1 \to \mathbb{R}$ is harmonic as well.

Now suppose that $\varphi : D_1 \to D_2$ is holomorphic, a bijection, and also that $\varphi^{-1} : D_2 \to D_1$ is holomorphic. We call such a map φ a *biholomorphism*. Conclude that a function $h : D_2 \to \mathbb{R}$ is harmonic if and only if $h \circ \varphi : D_1 \to \mathbb{R}$ is harmonic.

Thus the existence of a biholomorphism between two domains allows one to transplant harmonic (or even holomorphic) functions from one domain to the other. This mobility has the advantage that if D_1 is "nice" (like a half plane or a disc), while D_2 is complicated, then problems (like the Dirichlet Problem) in D_2 can be solved by first moving over to D_1, solving it there, and then transplanting the solution to D_2.

A first natural question is then the following: Given two domains D_1 and D_2, is there a biholomorphism between them? An answer is provided by the Riemann Mapping Theorem, a proof of which is beyond the scope of this book, but can be found for example in [Conway (1978)].

Theorem 5.5. (Riemann Mapping Theorem) *Let D be a proper (that is, $D \neq \mathbb{C}$) simply connected domain in $\mathbb{C}$. Then there exists a biholomorphism $\varphi : D \to \mathbb{D} := \{z \in \mathbb{C} : |z| < 1\}$.*

Thus the above result guarantees a biholomorphism between any two proper simply connected domains (by a passage through $\mathbb{D}$). Unfortunately, the proof does not give a practical algorithm for finding the biholomorphism. Show that the "Möbius transformation" $\varphi : \mathbb{H} \to \mathbb{D}$, where $\mathbb{H} := \{s \in \mathbb{C} : \mathrm{Re}(s) > 0\}$, given by

$$\varphi(s) = \frac{s - 1}{s + 1}, \quad s \in \mathbb{H},$$

is a biholomorphism between the right half plane $\mathbb{H}$ and the disc $\mathbb{D}$.

5.5 Notes

Exercises 5.5, 5.6 and the proof of Theorem 5.2 is taken from [Beck, Marchesi, Pixton, Sabalka (2008)]. Exercise 5.8 is based on [Flanigan (1973)].

Solutions

Solutions to the exercises from the Introduction

Solution to Exercise 0.1

Suppose the derivative of f' at 0 exists, and is equal to L. Then taking $\epsilon := 1 > 0$, there exists a $\delta > 0$ such that whenever $0 < |x - 0| < \delta$, we have

$$\left| \frac{f'(x) - f'(0)}{x - 0} - L \right| < \epsilon.$$

In particular, with $x := \delta/2$, we have $0 < |x - 0| = \delta/2 < \delta$, and so there must hold that

$$\left| \frac{f'(x) - f'(0)}{x - 0} - L \right| = \left| \frac{2(\delta/2) - 0}{(\delta/2) - 0} - L \right| = |2 - L| < \epsilon. \qquad (5.5)$$

On the other hand, with $x := -\delta/2$, we have $0 < |x - 0| = \delta/2 < \delta$, and so again there must hold that

$$\left| \frac{f'(x) - f'(0)}{x - 0} - L \right| = \left| \frac{-2(-\delta/2) - 0}{(-\delta/2) - 0} - L \right| = |2 + L| < \epsilon. \qquad (5.6)$$

From (5.5) and (5.6) it follows, using the triangle inequality for the real absolute value, that $4 = |2 + L + 2 - L| \leq |2 + L| + |2 - L| < \epsilon + \epsilon = 2\epsilon = 2$, a contradiction. Hence f' cannot be differentiable at 0.

Solutions to the exercises from Chapter 1

Solution to Exercise 1.1

Since $(x, y) \neq 0$, at least one among x, y is nonzero, and so $x^2 + y^2 \neq 0$. Thus

$$\left(\frac{x}{x^2 + y^2}, \frac{-y}{x^2 + y^2} \right) \in \mathbb{R}^2.$$

Moreover,

$$(x, y) \cdot \left(\frac{x}{x^2 + y^2}, \frac{-y}{x^2 + y^2} \right)$$

$$= \left(x \cdot \frac{x}{x^2 + y^2} - y \cdot \left(\frac{-y}{x^2 + y^2} \right), \ x \cdot \left(\frac{-y}{x^2 + y^2} \right) + y \cdot \frac{x}{x^2 + y^2} \right)$$

$$= \left(\frac{x^2 + y^2}{x^2 + y^2}, \frac{-xy + xy}{x^2 + y^2} \right) = (1, 0).$$

Hence for $(x, y) \neq (0, 0)$, we have $(x, y)^{-1} = \left(\dfrac{x}{x^2 + y^2}, \dfrac{-y}{x^2 + y^2} \right)$ in $\mathbb{C}$.

Solution to Exercise 1.2

Since $\theta \in \left(-\dfrac{\pi}{2}, \dfrac{\pi}{2} \right)$, $\tan \theta \in \mathbb{R}$. We have

$$\frac{1}{1 - i \tan \theta} = \frac{1}{1^2 + (\tan \theta)^2} + i \left(\frac{\tan \theta}{1^2 + (\tan \theta)^2} \right)$$

$$= \frac{(\cos \theta)^2}{(\cos \theta)^2 + (\sin \theta)^2} + i \left(\frac{\dfrac{\sin \theta}{\cos \theta} \cdot (\cos \theta)^2}{(\cos \theta)^2 + (\sin \theta)^2} \right)$$

$$= \frac{(\cos \theta)^2}{1} + i \frac{(\sin \theta)(\cos \theta)}{1} = (\cos \theta)^2 + i(\sin \theta)(\cos \theta).$$

Hence

$$\frac{1 + i \tan \theta}{1 - i \tan \theta} = (1 + i \tan \theta)((\cos \theta)^2 + i(\sin \theta)(\cos \theta))$$

$$= (\cos \theta)^2 - \frac{\sin \theta}{\cos \theta} \cdot (\sin \theta)(\cos \theta)$$

$$+ i \left((\sin \theta)(\cos \theta) + \frac{\sin \theta}{\cos \theta} \cdot (\cos \theta)^2 \right)$$

$$= (\cos \theta)^2 - (\sin \theta)^2 + i 2 (\sin \theta)(\cos \theta) = \cos(2\theta) + i \sin(2\theta).$$

Solution to Exercise 1.3

Let $P \subset \mathbb{C}$ be a set of positive elements of $\mathbb{C}$. Then since $i \neq 0$, by (P3), either $i \in P$ or $(i \notin P$ and $-i \in P)$. By (P2), we have

$$-1 = i \cdot i = (-i) \cdot (-i) \in P. \tag{5.7}$$

Again by (P2),

$$1 = (-1) \cdot (-1) \in P. \tag{5.8}$$

But $1 \neq 0$, and (5.7), (5.8) contradict (P3) for $x = 1$.

Solution to Exercise 1.4

See Figure 5.2.

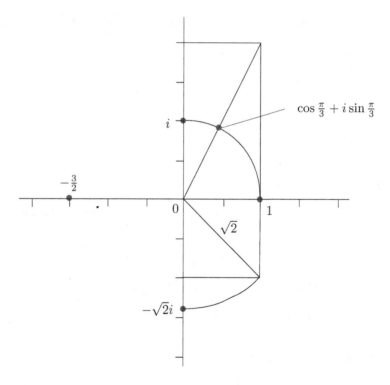

Fig. 5.2 Location of the complex numbers 0, 1, $-3/2$, i, $-\sqrt{2}i$, $\cos\dfrac{\pi}{3} + i\sin\dfrac{\pi}{3}$.

Solution to Exercise 1.5

We have for $\theta \in \mathbb{R}$ that $(\cos\theta + i\sin\theta)^3 = \cos(3\theta) + i\sin(3\theta)$. But

$$
\begin{aligned}
(\cos\theta + i\sin\theta)^3 &= (\cos\theta + i\sin\theta)((\cos\theta)^2 - (\sin\theta)^2 + i2(\cos\theta)(\sin\theta) \\
&= (\cos\theta)((\cos\theta)^2 - (\sin\theta)^2) - (\sin\theta)2(\cos\theta)(\sin\theta) \\
&\quad + i(\cdots).
\end{aligned}
$$

Hence equating the real parts on both sides, we obtain

$$
\begin{aligned}
\cos(3\theta) &= \operatorname{Re}((\cos\theta + i\sin\theta)^3) \\
&= (\cos\theta)((\cos\theta)^2 - (\sin\theta)^2) - 2(\cos\theta)(\sin\theta)^2 \\
&= (\cos\theta)((\cos\theta)^2 - 1 + (\cos\theta)^2) - 2(\cos\theta)(1 - (\cos\theta)^2) \\
&= (\cos\theta)^3 - \cos\theta + (\cos\theta)^3 - 2\cos\theta + 2(\cos\theta)^3 \\
&= 4(\cos\theta)^3 - 3\cos\theta.
\end{aligned}
$$

Alternatively, the Binomial Formula $(a+b)^n = \displaystyle\sum_{k=0}^{n} \binom{n}{k} a^k b^{n-k}$ continues to hold for $a, b \in \mathbb{C}$, $n \in \mathbb{N}$ and so

$$
\begin{aligned}
\cos(3\theta) &= \operatorname{Re}((\cos\theta + i\sin\theta)^3) \\
&= \operatorname{Re}((\cos\theta)^3 + 3(\cos\theta)^2(i\sin\theta) + 3(\cos\theta)(i\sin\theta)^2 + (i\sin\theta)^3) \\
&= (\cos\theta)^3 - 3(\cos\theta)(\sin\theta)^2 \\
&= 4(\cos\theta)^3 - 3\cos\theta.
\end{aligned}
$$

Solution to Exercise 1.6

We have $1 + i = \sqrt{2}\left(\dfrac{1}{\sqrt{2}} + i\dfrac{1}{\sqrt{2}}\right) = \sqrt{2}\left(\cos\dfrac{\pi}{4} + i\sin\dfrac{\pi}{4}\right)$. Hence

$$
\begin{aligned}
(1+i)^{10} &= (\sqrt{2})^{10}\left(\cos\frac{\pi}{4} + i\sin\frac{\pi}{4}\right)^{10} = 2^5\left(\cos\left(10 \cdot \frac{\pi}{4}\right) + i\sin\left(10 \cdot \frac{\pi}{4}\right)\right) \\
&= 32\left(\cos\left(2\pi + \frac{\pi}{2}\right) + i\sin\left(2\pi + \frac{\pi}{2}\right)\right) \\
&= 32\left(\cos\left(\frac{\pi}{2}\right) + i\sin\left(\frac{\pi}{2}\right)\right) = 32(0 + i \cdot 1) = 32i.
\end{aligned}
$$

Solution to Exercise 1.7

The angle made by $2 + i$ with the positive real axis is $\tan^{-1}(1/2)$, and the angle made by $3 + i$ with the positive real axis is $\tan^{-1}(1/3)$. Thus the angle made by $(2+i)(3+i)$ with the positive real axis is $\tan^{-1}(1/2)+\tan^{-1}(1/3)$. On the other hand, since

$$(2 + i)(3 + i) = 6 - 1 + i(2 + 3) = 5 + 5i,$$

the angle made by $(2 + i)(3 + i)$ with the positive real axis is

$$\tan^{-1}(5/5) = \tan^{-1} 1 = \pi/4.$$

Consequently, $\dfrac{\pi}{4} = \tan^{-1}\dfrac{1}{2} + \tan^{-1}\dfrac{1}{3}$.

Solution to Exercise 1.8

Suppose that the vertices A, B, C of the equilateral triangle are at the complex numbers z_A, z_B, z_C, and that they are labelled in the anticlockwise fashion. Since $\ell(AC) = \ell(AB)$ and $\angle CAB = \pi/3$, we have

$$z_C - z_A = \left(\cos\frac{\pi}{3} + i\sin\frac{\pi}{3}\right)(z_B - z_A). \tag{5.9}$$

We argue by contradiction and let $p, q, m, n \in \mathbb{Z}$ be such that

$$z_C - z_A = p + iq, \text{ and}$$
$$z_B - z_A = m + in.$$

Then (5.9) becomes $p + iq = \left(\dfrac{1}{2} + \dfrac{\sqrt{3}}{2}i\right)(m + in)$, that is,

$$p = \frac{m}{2} - \frac{\sqrt{3}}{2}n, \text{ and} \tag{5.10}$$

$$q = \frac{m\sqrt{3}}{2} + \frac{n}{2}. \tag{5.11}$$

Thus (by multiplying (5.10) by $-n$ and (5.11) by m and adding), we obtain

$$qm - pn = \frac{\sqrt{3}}{2}(m^2 + n^2).$$

But $m^2 + n^2 \neq 0$ (since $z_B \neq z_A$!), and so we obtain

$$\sqrt{3} = \frac{2(qm - pn)}{m^2 + n^2} \in \mathbb{Q},$$

a contradiction.

Solution to Exercise 1.9

We write $-1 = 1 \cdot (\cos \pi + i \sin \pi)$, and we seek $w = \rho(\cos \alpha + i \sin \alpha)$ such that $w^4 = \rho^4(\cos(4\alpha) + i \sin(4\alpha)) = 1 \cdot (\cos \pi + i \sin \pi)$. Hence $\rho^4 = 1$ and so $\rho = 1$. Also, we have $4\alpha \in \{\pi, \ \pi \pm 2\pi, \ \pi \pm 4\pi, \ \cdots\}$, and so

$$\alpha \in \left\{\frac{\pi}{4}, \ \frac{\pi}{4} \pm \frac{\pi}{2}, \ \frac{\pi}{4} \pm \pi, \ \cdots\right\}.$$

So we get $w = \rho(\cos \alpha + i \sin \alpha) = 1(\cos \alpha + i \sin \alpha)$ belongs to

$$\left\{\cos\frac{\pi}{4} + i\sin\frac{\pi}{4}, \ \cos\frac{3\pi}{4} + i\sin\frac{3\pi}{4}, \ \cos\frac{5\pi}{4} + i\sin\frac{5\pi}{4}, \ \cos\frac{7\pi}{4} + i\sin\frac{7\pi}{4}\right\}$$

$$= \left\{\frac{1+i}{\sqrt{2}}, \frac{-1+i}{\sqrt{2}}, \frac{-1-i}{\sqrt{2}}, \frac{1-i}{\sqrt{2}}\right\}.$$

The four fourth roots of unity are depicted in the complex plane in Figure 5.3.

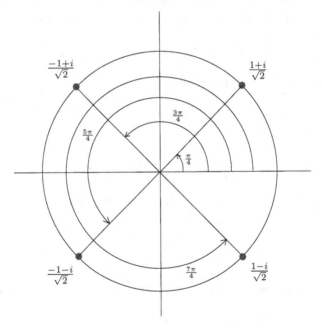

Fig. 5.3   Location of the complex numbers w that satisfy $w^4 = -1$.

Solution to Exercise 1.10

We have

$$0 = z^6 - z^3 - 2 = (z^3)^2 - 2z^3 + z^3 - 2 = (z^3 - 2)(z^3 + 1),$$

and so $z^3 = 2$ or $z^3 = -1$. The equation $z^3 = 2$ holds if and only if

$$z \in \left\{ \sqrt[3]{2} \left(\cos \frac{2\pi}{3} + i \sin \frac{2\pi}{3} \right), \ \sqrt[3]{2} \left(\cos \frac{4\pi}{3} + i \sin \frac{4\pi}{3} \right), \ \sqrt[3]{2} \right\},$$

that is,

$$z \in \left\{ \sqrt[3]{2} \left(-\frac{1}{2} + i \frac{\sqrt{3}}{2} \right), \ \sqrt[3]{2} \left(-\frac{1}{2} - i \frac{\sqrt{3}}{2} \right), \ \sqrt[3]{2} \right\}.$$

On the other hand, $z^3 = -1$ holds if and only if

$$z \in \left\{ \cos \frac{\pi}{3} + i \sin \frac{\pi}{3}, \ \cos \pi + i \sin \pi, \ \cos \frac{5\pi}{3} + i \sin \frac{5\pi}{3} \right\},$$

that is,

$$z \in \left\{ \frac{1}{2} + i \frac{\sqrt{3}}{2}, \ -1, \ \frac{1}{2} - i \frac{\sqrt{3}}{2} \right\}.$$

So $z^6 - z^3 - 2 = 0$ if and only if $[(z^3 = 2)$ or $(z^3 = -1)]$, that is, if and only if

$$z \in \left\{ \sqrt[3]{2} \left(-\frac{1}{2} + i \frac{\sqrt{3}}{2} \right), \ \sqrt[3]{2} \left(-\frac{1}{2} - i \frac{\sqrt{3}}{2} \right), \ \sqrt[3]{2} \right\}$$

$$\bigcup \left\{ \frac{1}{2} + i \frac{\sqrt{3}}{2}, \ -1, \ \frac{1}{2} - i \frac{\sqrt{3}}{2} \right\},$$

that is,

$$z \in \left\{ \sqrt[3]{2} \left(-\frac{1}{2} + i \frac{\sqrt{3}}{2} \right), \ \sqrt[3]{2} \left(-\frac{1}{2} - i \frac{\sqrt{3}}{2} \right), \ \sqrt[3]{2}, \frac{1}{2} + i \frac{\sqrt{3}}{2}, \ -1, \ \frac{1}{2} - i \frac{\sqrt{3}}{2} \right\}.$$

Solution to Exercise 1.11

Suppose $\omega \in \mathbb{C} \setminus \mathbb{R}$ is such that $\omega^3 = 1$. Then $(\omega - 1)(\omega^2 + \omega + 1) = 0$, and since $\omega \neq 1$, we have $\omega^2 + \omega + 1 = 0$. Hence

$$\begin{aligned}
((b - a)\omega &+ (b - c))((b - a)\omega^2 + b - c) \\
&= (b - a)^2 \omega^3 + (b - a)(b - c)(-1) + (b - c)^2 \\
&= (b - a)^2 \cdot 1 + (b - a)(b - c)(-1) + (b - c)^2 \\
&= (b - a)(b - a - b + c) + (b - c)^2 \\
&= (b - a)(c - a) + (b - c)^2 \\
&= (bc - ca - ab + a^2 + b^2 - 2bc + c^2) \\
&= a^2 + b^2 + c^2 - ab - bc - ca = 0.
\end{aligned}$$

Hence either $(b-a)\omega = c-b$ or $(b-a)\omega^2 = c-b$. But the latter case is the same as $(b-a)\omega^3 = (c-b)\omega$, that is, $(c-b)\omega = b-a$. So we have that $|b-a| = |c-b|$, and the angle between the line segments joining a to b and a to c is $\pi/3$; see Figure 5.4.

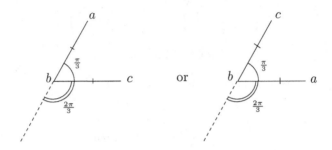

Fig. 5.4　a, b, c form an equilateral triangle.

In either case, we obtain that the triangle formed by a, b, c is equilateral.

If a, b, c are all real, then the equilateral triangle must degenerate to a point $r \in \mathbb{R}$, and so $a = b = c \; (= r)$. Thus we recover the real case result.

Solution to Exercise 1.12

Let $\omega \in \mathbb{C} \backslash \mathbb{R}$ be such that $\omega^3 = 1$. Since $(1-\omega)(1+\omega+\omega^2) = 0$, and $\omega \neq 1$, we have $1+\omega+\omega^2 = 0$. Also, $1+\omega^2+\omega^4 = 1+\omega^2+\omega \cdot \omega^3 = 1+\omega^2+\omega = 0$. We have

$$(1+1)^{3n} + (1+\omega)^{3n} + (1+\omega^2)^{3n} = \sum_{k=0}^{3n} \binom{3n}{k} (1+\omega^k + \omega^{2k}).$$

But

$$(1+\omega^k + \omega^{2k}) = \begin{cases} 1+1+1 & \text{if } k \equiv 0 \mod 3, \\ 1+\omega+\omega^2 & \text{if } k \equiv 1 \mod 3, \\ 1+\omega^2+\omega^4 & \text{if } k \equiv 2 \mod 3 \end{cases}$$

$$= \begin{cases} 3 & \text{if } k \equiv 0 \mod 3, \\ 0 & \text{if } k \equiv 1 \mod 3, \\ 0 & \text{if } k \equiv 2 \mod 3. \end{cases}$$

Thus

$$(1+1)^{3n} + (1+\omega)^{3n} + (1+\omega^2)^{3n} = 3 \cdot \left(\binom{3n}{0} + \binom{3n}{3} + \cdots + \binom{3n}{3n} \right).$$

But also

$$(1+1)^{3n} + (1+\omega)^{3n} + (1+\omega^2)^{3n} = 2^{3n} + (-\omega^2)^{3n} + (-\omega)^{3n}$$
$$= 2^{3n} + (-1)^n + (-1)^n$$
$$= 2^{3n} + 2(-1)^n,$$

and so the claim follows.

Solution to Exercise 1.13

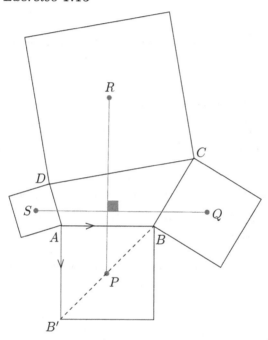

Fig. 5.5 RP and SQ have equal lengths and meet at right angles.

See Figure 5.5. Let the points A, B, C, D in the plane correspond to the complex numbers a, b, c, d, respectively. Since AB' is obtained from AB by rotating AB about A in an anticlockwise fashion by $90°$, we have that B' corresponds to the complex number $a - i(b - a)$. Since P is the midpoint of BB', it follows that P corresponds to

$$\frac{a + b - i(b - a)}{2}.$$

Similarly, Q, R, S correspond to

$$\frac{b + c - i(c - b)}{2}, \quad \frac{c + d - i(d - c)}{2}, \quad \frac{d + a - i(a - d)}{2},$$

respectively. If we denote the complex numbers corresponding to P, Q, R, S, by p, q, r, s, respectively, then

$$i(q - s) = i\left(\frac{b + c - i(c - b)}{2} - \frac{d + a - i(a - d)}{2}\right)$$

$$= \frac{-b + c - a + d + i(b + c - d - a)}{2}$$

$$= \frac{-a - b + i(b - a)}{2} + \frac{c + d - i(d - c)}{2} = -p + r.$$

Hence $|q - s| = |p - r|$, showing that $\ell(QS) = \ell(PR)$. Also, since multiplication by i produces a rotation about the origin by $90°$, we see that $PR \perp QS$.

Solution to Exercise 1.14

Let $z_1 = x_1 + iy_1$, $z_2 = x_2 + iy_2$, where x_1, x_2, y_1, y_2 belong to $\mathbb{R}$. Then $z_1 z_2 = x_1 x_2 - y_1 y_2 + i(x_1 y_2 + y_1 x_2)$, and

$$|z_1 z_2|^2 = (x_1 x_2 - y_1 y_2)^2 + (x_1 y_2 + y_1 x_2)^2$$

$$= x_1^2 x_2^2 - 2x_1 x_2 y_1 y_2 + y_1^2 y_2^2 + x_1^2 y_2^2 + 2x_1 y_2 y_1 x_2 + y_1^2 x_2^2$$

$$= x_1^2(x_2^2 + y_2^2) + y_1^2(y_2^2 + x_2^2) = (x_1^2 + y_1^2)(x_2^2 + y_2^2)$$

$$= |z_1|^2 |z_2|^2.$$

Since $|z_1|, |z_2|, |z_1 z_2|$ are all nonnegative, it follows that $|z_1 z_2| = |z_1| |z_2|$.

Solution to Exercise 1.15

Let $z = x + iy$, where $x, y \in \mathbb{R}$. Then

$$\overline{(\overline{z})} = \overline{x - iy} = x - i(-y) = x + iy = z.$$

Also,

$$z\overline{z} = (x + iy)(x - iy) = x^2 + y^2 + i(-xy + xy) = x^2 + y^2 = |z|^2.$$

Finally,

$$\frac{z + \overline{z}}{2} = \frac{x + iy + x - iy}{2} = \frac{2x}{2} = x = \text{Re}(z), \text{ and}$$

$$\frac{z - \overline{z}}{2i} = \frac{x + iy - x + iy}{2i} = \frac{2iy}{2i} = y = \text{Im}(z).$$

Solution to Exercise 1.16

Let $z = x + iy$, where $x, y \in \mathbb{R}$. Then

$$|z| = |x + iy| = \sqrt{x^2 + y^2} = \sqrt{x^2 + (-y)^2} = |x - iy| = |\bar{z}|,$$
$$|\mathrm{Re}(z)| = |x| = \sqrt{x^2} \le \sqrt{x^2 + y^2} = |x + iy| = |z|,$$
$$|\mathrm{Im}(z)| = |y| = \sqrt{y^2} \le \sqrt{x^2 + y^2} = |x + iy| = |z|.$$

Since $\bar{z}$ is obtained by reflection of z in the real axis and since $0 \in \mathbb{R}$, the distance of z to 0 is equal to the distance of $\bar{z}$ to 0. In other words, $|z| = |\bar{z}|$. The inequalities $|\mathrm{Re}(z)| \le |z|$ and $|\mathrm{Im}(z)| \le |z|$ just say that the length of any side in a right angled triangle is at most the length of the hypotenuse.

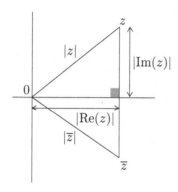

Solution to Exercise 1.17

First we note that $|\bar{a}z| = |\bar{a}| \, |z| = |a| \, |z| < 1 \cdot 1 = 1$, and so $\bar{a}z \ne 1$. We have

$$\frac{z - a}{1 - \bar{a}z} \cdot \overline{\left(\frac{z - a}{1 - \bar{a}z}\right)} = \frac{z - a}{1 - \bar{a}z} \cdot \frac{\bar{z} - \bar{a}}{1 - a\bar{z}} = \frac{z\bar{z} - a\bar{z} - \bar{a}z + a\bar{a}}{1 - a\bar{z} - \bar{a}z + a\bar{a}z\bar{z}}$$

$$= \frac{|z|^2 - a\bar{z} - \bar{a}z + |a|^2}{1 - a\bar{z} - \bar{a}z + |a|^2|z|^2}$$

$$= \frac{1 - a\bar{z} - \bar{a}z + |a|^2|z|^2 + |z|^2 + |a|^2 - 1 - |a|^2|z|^2}{1 - a\bar{z} - \bar{a}z + |a|^2|z|^2}$$

$$= 1 + \frac{|z|^2 + |a|^2 - 1 - |a|^2|z|^2}{1 - a\bar{z} - \bar{a}z + |a|^2|z|^2}$$

$$= 1 + \frac{|z|^2 + |a|^2 - 1 - |a|^2|z|^2}{|1 - \bar{a}z|^2}$$

$$= 1 - \frac{(1 - |z|^2)(1 - |a|^2)}{|1 - \bar{a}z|^2}.$$

Thus $\left| \dfrac{z-a}{1-\overline{a}z} \right|^2 = 1 - \underbrace{\dfrac{(1-|z|^2)(1-|a|^2)}{|1-\overline{a}z|^2}}_{\geq 0 \text{ as } |z| \leq 1, \ |a| < 1} \leq 1 - 0 = 1.$

Solution to Exercise 1.18

Let $w \in \mathbb{C}$ be such that $p(w) = 0$, that is, $c_0 + c_1 w + \cdots + c_d w^d = 0$. Hence

$$\overline{c_0 + c_1 w + \cdots + c_d w^d} = \overline{0} = 0,$$

and so, using the fact that the c_k are all real for $0 \leq k \leq d$, we obtain

$$0 = \overline{c_0 + c_1 w + \cdots + c_d w^d} = \overline{c_0} + \overline{c_1 w} + \cdots + \overline{c_d w^d}$$
$$= c_0 + \overline{c_1}\,\overline{w} + \cdots + \overline{c_d}\,\overline{w^d} = c_0 + c_1 \overline{w} + \cdots + c_d (\overline{w})^d$$

where the last equality follows from the observation that

$$\overline{w^k} = \overline{\underbrace{w \cdots w}_{k \text{ times}}} = \underbrace{\overline{w} \cdots \overline{w}}_{k \text{ times}} = (\overline{w})^k, \quad 1 \leq k \leq d.$$

Consequently, $0 = c_0 + c_1 \overline{w} + \cdots + c_d (\overline{w})^d = p(\overline{w})$.

Solution to Exercise 1.19

Let $a = |a|(\cos\alpha + i\sin\alpha)$, and $b = |b|(\cos\beta + i\sin\beta)$, where $\alpha, \beta \in [0, 2\pi)$. Then

$$a\overline{b} = |a|(\cos\alpha + i\sin\alpha) \cdot |b|(\cos\beta - i\sin\beta)$$
$$= |a|\,|b|(\cos\alpha + i\sin\alpha)(\cos\beta - i\sin\beta),$$

and so $\text{Im}(a\overline{b}) = |a|\,|b|(-(\cos\alpha)(\sin\beta) + (\sin\alpha)(\cos\beta)) = |a|\,|b|\sin(\alpha - \beta).$

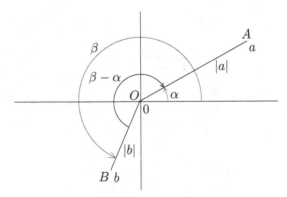

Fig. 5.6 The area of $\triangle OAB$ formed by the triangle with vertices at 0, a, b.

The area of the triangle OAB formed by 0, a, b (where $O \equiv 0$, $A \equiv a$ and $B \equiv b$) is

$$\frac{1}{2}\ell(OA)\cdot\ell(OB)\cdot\sin \angle AOB = \frac{1}{2}|a|\cdot|b|\cdot|\sin(\alpha-\beta)| = \frac{1}{2}|\mathrm{Im}(a\bar{b})| = \left|\frac{\mathrm{Im}(a\bar{b})}{2}\right|.$$

Solution to Exercise 1.20

For $z_1, z_2, z_3 \in \mathbb{C}$, we have

$$w := i \cdot \overline{\det \begin{bmatrix} 1 & z_1 & \overline{z_1} \\ 1 & z_2 & \overline{z_2} \\ 1 & z_3 & \overline{z_3} \end{bmatrix}} = -i \cdot \overline{\det \begin{bmatrix} 1 & z_1 & \overline{z_1} \\ 1 & z_2 & \overline{z_2} \\ 1 & z_3 & \overline{z_3} \end{bmatrix}}.$$

But for a square matrix $M = [m_{ij}]$,

$$\det M = \sum_{\sigma \in S_n} (\mathrm{sign}\ \sigma) \cdot m_{i\sigma(i)},$$

where S_n is the set of all permutations on the set $\{1, \cdots, n\}$. Hence

$$\overline{\det M} = \sum_{\sigma \in S_n} (\mathrm{sign}\ \sigma) \cdot \overline{m_{i\sigma(i)}} = \det \overline{M},$$

where $\overline{M}$ is the matrix obtained from M by taking entrywise complex conjugates. Hence

$$\overline{\det \begin{bmatrix} 1 & z_1 & \overline{z_1} \\ 1 & z_2 & \overline{z_2} \\ 1 & z_3 & \overline{z_3} \end{bmatrix}} = \det \begin{bmatrix} 1 & \overline{z_1} & z_1 \\ 1 & \overline{z_2} & z_2 \\ 1 & \overline{z_3} & z_3 \end{bmatrix} = -\det \begin{bmatrix} 1 & z_1 & \overline{z_1} \\ 1 & z_2 & \overline{z_2} \\ 1 & z_3 & \overline{z_3} \end{bmatrix},$$

where the last equality is obtained by interchanging the second and the third columns. Consequently,

$$i \cdot \overline{\det \begin{bmatrix} 1 & z_1 & \overline{z_1} \\ 1 & z_2 & \overline{z_2} \\ 1 & z_3 & \overline{z_3} \end{bmatrix}} = -i \cdot \overline{\det \begin{bmatrix} 1 & z_1 & \overline{z_1} \\ 1 & z_2 & \overline{z_2} \\ 1 & z_3 & \overline{z_3} \end{bmatrix}} = -i \cdot \left(-\det \begin{bmatrix} 1 & z_1 & \overline{z_1} \\ 1 & z_2 & \overline{z_2} \\ 1 & z_3 & \overline{z_3} \end{bmatrix} \right)$$

$$= i \cdot \det \begin{bmatrix} 1 & z_1 & \overline{z_1} \\ 1 & z_2 & \overline{z_2} \\ 1 & z_3 & \overline{z_3} \end{bmatrix}.$$

Thus w is its own complex conjugate, and hence it is real.

Solution to Exercise 1.21

We have

$$|z_1 + z_2|^2 + |z_1 - z_2|^2$$

$$= (z_1 + z_2)(\overline{z_1} + \overline{z_2}) + (z_1 - z_2)(\overline{z_1} - \overline{z_2})$$

$$= z_1 \cdot \overline{z_1} + z_1 \cdot \overline{z_2} + z_2 \cdot \overline{z_1} + z_2 \cdot \overline{z_2}$$

$$+ z_1 \cdot \overline{z_1} + z_1 \cdot (-\overline{z_2}) + (-z_2) \cdot \overline{z_1} + (-z_2) \cdot (-\overline{z_2})$$

$$= |z_1|^2 + \cancel{z_1 \, \overline{z_2}} + \cancel{z_2 \, \overline{z_1}} + |z_2|^2 + |z_1|^2 - \cancel{z_1 \, \overline{z_2}} - \cancel{z_2 \, \overline{z_1}} + |z_2|^2$$

$$= 2(|z_1|^2 + |z_2|^2).$$

Consider the paralellogram P with vertices at $0, z_1, z_2, z_1 + z_2$ in the complex plane. Then $|z_1 + z_2|$ denotes the length of one diagonal of P, while $|z_1 - z_2|$ is the length of the other diagonal of P. Also, $|z_1|, |z_2|$ are the lengths of the sides on P. So the above equality says "In a parallelogram, the sum of the squares of the lengths of the diagonals equals twice the sum of the squares of the sides."

Solution to Exercise 1.22

For $z_1, z_2 \in \mathbb{C}$, we have $|z_1| = |z_1 - z_2 + z_2| \leq |z_1 - z_2| + |z_2|$, and so

$$|z_1| - |z_2| \leq |z_1 - z_2|. \tag{5.12}$$

Since this holds for *all* $z_1, z_2 \in \mathbb{C}$, by swapping z_1 and z_2 in (5.12),

$$|z_2| - |z_1| \leq |z_2 - z_1| = |-(z_1 - z_2)| = |-1||z_1 - z_2| = |z_1 - z_2|. \tag{5.13}$$

(5.12) and (5.13) give $||z_1| - |z_2|| \leq |z_1 - z_2|$.

Solution to Exercise 1.23

(1),(2),(3):

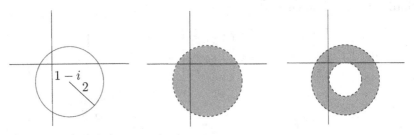

Fig. 5.7   Left to right: The set of points described by $|z - (1 - i)| = 2$, $|z - (1 - i)| < 2$ and $1 < |z - (1 - i)| < 2$, respectively.

(4) Let $z = x + iy$, where $x, y \in \mathbb{R}$. Then $\text{Re}(z - (1 - i)) = 3$ if and only if $x - 1 = 3$, that is $x = 4$.

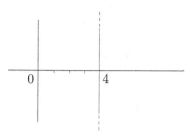

(5) Let $z = x + iy$, where $x, y \in \mathbb{R}$. Then $|\text{Im}(z - (1 - i))| < 3$ if and only if $|y + 1| < 3$, that is $-4 < y < 2$.

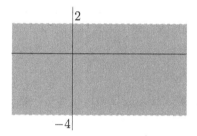

(6) $\{z \in \mathbb{C} : |z - (1 - i)| = |z - (1 + i)|\}$ is the set of all complex numbers z whose distance to $1 - i$ is equal to its distance to $1 + i$. So it is the set of all points which lies on the perpendicular bisector of the line segment joining $1 - i$ to $1 + i$. So the set is the real line $\mathbb{R}$.

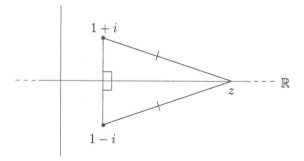

Fig. 5.8 The set of points z satisfying $|z - (1 - i)| = |z - (1 + i)|$ is $\mathbb{R}$.

(7) The equation $|z - (1 - i)| + |z - (1 + i)| = 2$ says that the sum of the

distances of z to $1+i$ and to $1-i$ is 2. But the distance between $1-i$ and $1+i$ is 2. So z lies on the line segment joining $1-i$ to $1-i$.

This conclusion can also be arrived at analytically. If $z = x + iy$, where $x, y \in \mathbb{R}$, then

$$2 = \sqrt{(x-1)^2 + (y+1)^2} + \sqrt{(x-1)^2 + (y-1)^2}$$
$$\geq |y+1| + |y-1| \geq 1 + \cancel{y} + 1 - \cancel{y} = 2.$$

Thus $|y+1| + |y-1| = 2$ and $x = 1$.

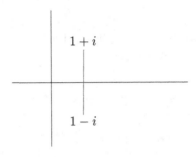

Fig. 5.9   The set of points z satisfying $|z - (1-i)| + |z - (1+i)| = 2$ is the line segment joining $1-i$ to $1+i$.

(8) The set of z such that $|z - (1-i)| + |z - (1+i)| = 3$ lies on an ellipse E with foci at $1+i$ and $1-i$, and so $\{z \in \mathbb{C} : |z - (1-i)| + |z - (1+i)| < 3\}$ is the region in the interior of the ellipse E.

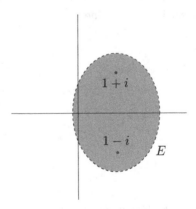

Fig. 5.10   The set of points z satisfying $|z - (1-i)| + |z - (1+i)| < 3$ is the interior of the ellipse E.

Solution to Exercise 1.24

For $z \neq 0$, we have $p(z) = z^d \left(c_d + \dfrac{c_{d-1}}{z} + \cdots + \dfrac{c_1}{z^{d-1}} + \dfrac{c_0}{z^d} \right)$. Since

$$\lim_{n \to \infty} \left(\frac{|c_{d-1}|}{n} + \cdots + \frac{|c_1|}{n^{d-1}} + \frac{|c_0|}{n^d} \right) = 0,$$

there exists an N large enough so that

$$\frac{|c_{d-1}|}{N} + \cdots + \frac{|c_1|}{N^{d-1}} + \frac{|c_0|}{N^d} < \frac{|c_d|}{2}.$$

Hence for $|z| > N =: R$, we have

$$|p(z)| = |z^d| \left| c_d + \frac{c_{d-1}}{z} + \cdots + \frac{c_1}{z^{d-1}} + \frac{c_0}{z^d} \right|$$

$$\geq |z|^d \left(|c_d| - \left| \frac{c_{d-1}}{z} + \cdots + \frac{c_1}{z^{d-1}} + \frac{c_0}{z^d} \right| \right)$$

$$\geq |z|^d \left(|c_d| - \left(\frac{|c_{d-1}|}{z} + \cdots + \frac{|c_1|}{z^{d-1}} + \frac{|c_0|}{z^d} \right) \right)$$

$$\geq |z|^d \left(|c_d| - \left(\frac{|c_{d-1}|}{N} + \cdots + \frac{|c_1|}{N^{d-1}} + \frac{|c_0|}{N^d} \right) \right)$$

$$\geq |z|^d \left(|c_d| - \frac{|c_d|}{2} \right) = \underbrace{\frac{|c_d|}{2}}_{=:M} |z|^d.$$

Solution to Exercise 1.25

("If" part) Suppose that the two real sequences $(\mathrm{Re}(z_n))_{n \in \mathbb{N}}$ and $(\mathrm{Im}(z_n))_{n \in \mathbb{N}}$ are convergent respectively to $\mathrm{Re}(L)$ and $\mathrm{Im}(L)$. Then given $\epsilon > 0$,

$$|z_n - L| = \sqrt{(\mathrm{Re}(z_n) - \mathrm{Re}(L))^2 + (\mathrm{Im}(z_n) - \mathrm{Im}(L))^2}$$

$$< \sqrt{\left(\frac{\epsilon}{\sqrt{2}} \right)^2 + \left(\frac{\epsilon}{\sqrt{2}} \right)^2} = \epsilon$$

for $n > N$, where N is large enough to ensure that for all $n > N$,

$$|\mathrm{Re}(z_n) - \mathrm{Re}(L)| < \frac{\epsilon}{\sqrt{2}} \text{ and } |\mathrm{Im}(z_n) - \mathrm{Im}(L)| < \frac{\epsilon}{\sqrt{2}}.$$

So $(z_n)_{n \in \mathbb{N}}$ converges to L.

("Only if" part) Suppose $(z_n)_{n \in \mathbb{N}}$ converges to L. Given $\epsilon > 0$, let N be such that for all $n > N$, $|z - L| < \epsilon$. But then for $n > N$,

$$|\mathrm{Re}(z_n) - \mathrm{Re}(L)| = |\mathrm{Re}(z_n - L)| \leq |z_n - L| < \epsilon, \text{ and}$$

$$|\mathrm{Im}(z_n) - \mathrm{Im}(L)| = |\mathrm{Im}(z_n - L)| \leq |z_n - L| < \epsilon.$$

Hence the two real sequences $(\mathrm{Re}(z_n))_{n \in \mathbb{N}}$ and $(\mathrm{Im}(z_n))_{n \in \mathbb{N}}$ are convergent respectively to $\mathrm{Re}(L)$ and $\mathrm{Im}(L)$.

Solution to Exercise 1.26

("Only if" part) Suppose $(z_n)_{n\in\mathbb{N}}$ converges to L. Then $(\mathrm{Re}(z_n))_{n\in\mathbb{N}}$ and $(\mathrm{Im}(z_n))_{n\in\mathbb{N}}$ converge to $\mathrm{Re}(L)$ and $\mathrm{Im}(L)$, respectively. Hence $(\mathrm{Re}(z_n))_{n\in\mathbb{N}}$ and $(-\mathrm{Im}(z_n))_{n\in\mathbb{N}}$ converge to $\mathrm{Re}(L)$ and $-\mathrm{Im}(L)$, respectively, that is, $(\mathrm{Re}(\overline{z_n}))_{n\in\mathbb{N}}$ and $(\mathrm{Im}(\overline{z_n}))_{n\in\mathbb{N}}$ converge to $\mathrm{Re}(\overline{L})$ and $\mathrm{Im}(\overline{L})$, respectively. Consequently $(\overline{z_n})_{n\in\mathbb{N}}$ converges to $\overline{L}$.

("If" part) Suppose $(\overline{z_n})_{n\in\mathbb{N}}$ converges to $\overline{L}$. By the previous part, $((\overline{\overline{z_n}}))_{n\in\mathbb{N}}$ converges to $(\overline{\overline{L}})$, that is, $(z_n)_{n\in\mathbb{N}}$ converges to L.   .

Solution to Exercise 1.27

Let $(z_n)_{n\in\mathbb{N}}$ be a Cauchy sequence in $\mathbb{C}$. The inequalities

$$|\mathrm{Re}(z_n) - \mathrm{Re}(z_m)| = |\mathrm{Re}(z_n - z_m)| \leq |z_n - z_m| \text{ and}$$
$$|\mathrm{Im}(z_n) - \mathrm{Im}(z_m)| = |\mathrm{Im}(z_n - z_m)| \leq |z_n - z_m|$$

show that the two real sequences $(\mathrm{Re}(z_n))_{n\in\mathbb{N}}$ and $(\mathrm{Im}(z_n))_{n\in\mathbb{N}}$ are then also Cauchy sequences. Since $\mathbb{R}$ is complete, they are convergent to, say, $a, b \in \mathbb{R}$, respectively. But then $(z_n)_{n\in\mathbb{N}}$ converges in $\mathbb{C}$ to $a + ib$. Hence $\mathbb{C}$ is complete.

Solution to Exercise 1.28

Let $z_0 \in \mathbb{C}$ and $\epsilon > 0$. Set $\delta = \epsilon > 0$. Then whenever $|z - z_0| < \delta$, we have

$$|\mathrm{Re}(z) - \mathrm{Re}(z_0)| = |\mathrm{Re}(z - z_0)| \leq |z - z_0| < \delta = \epsilon.$$

So $z \mapsto \mathrm{Re}(z)$ is continuous at z_0. Since the choice of $z_0 \in \mathbb{C}$ was arbitrary, it follows that $z \mapsto \mathrm{Re}(z)$ is continuous (in $\mathbb{C}$).

Solution to Exercise 1.29

Let $U := \{z \in \mathbb{C} : \mathrm{Re}(z) \cdot \mathrm{Im}(z) > 1\}$, and set $F := \complement U$ (the complement of U). If $(z_n)_{n\in\mathbb{N}}$ is a sequence in F such that $(z_n)_{n\in\mathbb{N}}$ converges to L in $\mathbb{C}$, then we have

$$\mathrm{Re}(z_n) \cdot \mathrm{Im}(z_n) \leq 1 \text{ for all } n \in \mathbb{N}, \tag{5.14}$$

and $(\mathrm{Re}(z_n))_{n\in\mathbb{N}}$, $(\mathrm{Im}(z_n))_{n\in\mathbb{N}}$ converge respectively to $\mathrm{Re}(L)$ and $\mathrm{Im}(L)$. Thus $(\mathrm{Re}(z_n) \cdot \mathrm{Im}(z_n))_{n\in\mathbb{N}}$ is also convergent, with limit $\mathrm{Re}(L) \cdot \mathrm{Im}(L)$. (5.14) then gives $\mathrm{Re}(L) \cdot \mathrm{Im}(L) \leq 1$. So $L \in F$. Hence F is closed, and so $\complement F = U$ is open.

Next we show that U is not a domain. Suppose that it is. Then there is a (stepwise) path $\gamma : [a, b] \to U$ that joins $\gamma(a) = 2 + 2i \in U$ to the point $\gamma(b) = -2 - 2i \in U$. Since the map $z \mapsto \operatorname{Re}(z) : \mathbb{C} \to \mathbb{R}$ is continuous, it follows that $t \overset{\varphi}{\longmapsto} \operatorname{Re}(\gamma(t)) : [a, b] \to \mathbb{R}$ is continuous too. We have

$$\varphi(a) = \operatorname{Re}(\gamma(a)) = \operatorname{Re}(2 + 2i) = 2, \text{ and}$$
$$\varphi(b) = \operatorname{Re}(\gamma(b)) = \operatorname{Re}(-2 - 2i) = -2.$$

Since $\varphi(a) = 2 > 0 > -2 = \varphi(b)$, it follows by the Intermediate Value Theorem that there exists a $t_* \in [a, b]$ such that $0 = \varphi(t_*) = \operatorname{Re}(\gamma(t_*))$. But then $\operatorname{Re}(\gamma(t_*)) \cdot \operatorname{Im}(\gamma(t_*)) = 0 \cdot \operatorname{Im}(\gamma(t_*)) = 0 \not> 1$, showing that $\gamma(t_*) \notin U$, a contradiction. So U is not path-connected, and hence not a domain.

Solution to Exercise 1.30

Since D is open, it is clear that its reflection in the real axis, D^*, is also open.

Let $w_1, w_2 \in D^*$. Then $\overline{w_1}, \overline{w_2} \in D$. As D is a domain, there exists a stepwise path $\gamma : [a, b] \to \mathbb{C}$ such that $\gamma(a) = \overline{w_1}$, $\gamma(b) = \overline{w_2}$ and for all $t \in [a, b]$, $\gamma(t) \in D$. Now define $\gamma^* : [a, b] \to \mathbb{C}$ by $\gamma^*(t) = \overline{\gamma(t)}$, $t \in [a, b]$. Then $\gamma^*(a) = \overline{\overline{w_1}} = w_1$, $\gamma^*(b) = \overline{\overline{w_2}} = w_2$, and for all $t \in [a, b]$, $\gamma^*(t) = \overline{\gamma(t)} \in D^*$. As γ^* is the composition of the continuous functions γ and $z \mapsto \bar{z}$, γ^* is continuous. Also, since γ is a stepwise path, there exist points

$$t_0 = a < t_1 < \cdots < t_n < t_{n+1} = b$$

such that for each $k = 0, 1, \cdots, n$, the restriction $\gamma|_{[t_k, t_{k+1}]}$ has either constant real part or constant imaginary part. Consequently $\gamma^*|_{[t_k, t_{k+1}]}$ (which has the same real part as $\gamma|_{[t_k, t_{k+1}]}$ and has imaginary part which is minus the imaginary part of $\gamma|_{[t_k, t_{k+1}]}$) also has either a constant real part of a constant imaginary part. So γ^* is a stepwise path too. Hence D^* is path-connected.

As D^* is open and path-connected, it is a domain.

Solution to Exercise 1.31

We have

$$\exp\left(i\frac{9\pi}{2}\right) = \exp\left(i\left(4\pi + \frac{\pi}{2}\right)\right) = e^0\left(\cos\frac{\pi}{2} + i\sin\frac{\pi}{2}\right) = 1(0 + i \cdot 1) = i,$$

and $\exp(3 + \pi i) = e^3(\cos\pi + i\sin\pi) = e^3(-1 + i \cdot 0) = -e^3$.

Solution to Exercise 1.32

Let $z = x + iy$, where $x, y \in \mathbb{R}$. Then $e^x(\cos y + i \sin y) = \pi i$. Taking absolute values on both sides, we obtain $e^x = \pi$, and so $x = \log \pi$. Hence $\cos y + i \sin y = i$, which means that $\sin y = 1$ and $\cos y = 0$. Thus $y = \frac{\pi}{2} + 2\pi k$, $k \in \mathbb{Z}$. See Figure 5.11.

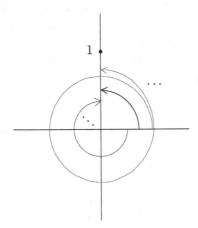

Fig. 5.11   Possible values of y when $\cos y + i \sin y = i$ are given by $y = \frac{\pi}{2} + 2\pi k$, $k \in \mathbb{Z}$.

Hence if $\exp z = \pi i$, then

$$z \in \left\{ \log \pi + i \left(\frac{\pi}{2} + 2\pi k \right), \ k \in \mathbb{Z} \right\}.$$

Vice versa, if $z \in \log \pi + i \left(\frac{\pi}{2} + 2\pi k \right)$ for some $k \in \mathbb{Z}$, then

$$\exp z = e^{\log \pi} \left(\cos \left(\frac{\pi}{2} + 2\pi k \right) + i \sin \left(\frac{\pi}{2} + 2\pi k \right) \right) = \pi(0 + i \cdot 1) = \pi i.$$

Consequently, $\exp z = \pi i$ if and only if $z \in \left\{ \log \pi + i \left(\frac{\pi}{2} + 2\pi k \right), \ k \in \mathbb{Z} \right\}.$

Solution to Exercise 1.33

Let $\gamma(t) := \exp(it)$, $t \in [0, 2\pi]$. Then

$$\gamma(t) = \exp(it) = e^0(\cos t + i \sin t) = \cos t + i \sin t.$$

The point $(\cos t, \sin t)$ lies on a circle of radius 1 and center $(0,0)$, and with increasing t, this point moves anticlockwise. Hence the curve $t \mapsto \gamma(t)$ is the circle traversed in the anticlockwise direction, as shown in Figure 5.12.

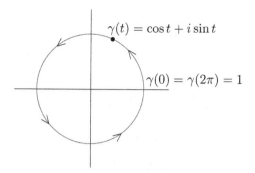

Fig. 5.12   The curve $t \mapsto \gamma(t) := \exp(it)$, $t \in [0, 2\pi]$.

Solution to Exercise 1.34

We have $\exp(t + it) = e^t(\cos t + i \sin t)$. So the image curve is given by $t \mapsto (e^t \cos t, e^t \sin t)$. We have sketched this (not to scale!) in Figure 5.13. Thus the curve is a spiral, and as $t \searrow -\infty$, $e^t(\cos t + i \sin t)$ converges to 0, while the curve spirals outwards as $t \nearrow +\infty$.

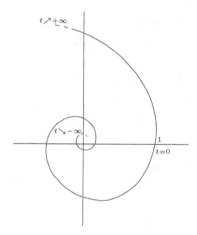

Fig. 5.13   The image of the line $y = x$ under the map $z = x + iy \mapsto \exp z$.

Solution to Exercise 1.35

We have

$$\exp(z^2) = \exp((x+iy)^2) = \exp(x^2 - y^2 + 2xyi) = e^{x^2 - y^2}(\cos(2xy) + i \sin(2xy)),$$

and so $|\exp(z^2)| = e^{x^2-y^2}$, $\mathrm{Re}(\exp(z^2)) = e^{x^2-y^2}\cos(2xy)$, and $\mathrm{Im}(\exp(z^2)) = e^{x^2-y^2}\sin(2xy)$. Also, we have for $z \neq 0$ that

$$\exp\frac{1}{z} = \exp\left(\frac{1}{x+iy}\right) = \exp\left(\frac{x-iy}{x^2+y^2}\right)$$

$$= e^{\frac{x}{x^2+y^2}}\left(\cos\left(\frac{-y}{x^2+y^2}\right) + i\sin\left(\frac{-y}{x^2+y^2}\right)\right),$$

and so it follows from here that

$$\left|\exp\frac{1}{z}\right| = e^{\frac{x}{x^2+y^2}},$$

$$\mathrm{Re}\left(\exp\frac{1}{z}\right) = e^{\frac{x}{x^2+y^2}}\cos\left(\frac{-y}{x^2+y^2}\right),$$

$$\mathrm{Im}\left(\exp\frac{1}{z}\right) = e^{\frac{x}{x^2+y^2}}\sin\left(\frac{-y}{x^2+y^2}\right).$$

Solution to Exercise 1.36

For $z_1, z_2 \in \mathbb{C}$, we have

$$(\sin z_1)(\cos z_2) + (\cos z_1)(\sin z_2)$$

$$= \left(\frac{\exp(iz_1) - \exp(-iz_1)}{2i}\right)\left(\frac{\exp(iz_2) + \exp(-iz_2)}{2}\right)$$

$$+ \left(\frac{\exp(iz_1) + \exp(-iz_2)}{2}\right)\left(\frac{\exp(iz_2) - \exp(-iz_2)}{2i}\right)$$

$$= \frac{2\exp(i(z_1+z_2)) - 2\exp(-i(z_1+z_2))}{4i} = \sin(z_1+z_2).$$

Solution to Exercise 1.37

Let $z = x + iy$, where $x, y \in \mathbb{R}$. Then

$$\cos z = \cos(x+iy) = (\cos x)(\cos(iy)) - (\sin x)(\sin(iy))$$

$$= (\cos x)\left(\frac{e^{-y}+e^y}{2}\right) - (\sin x)\left(\frac{e^{-y}-e^y}{2i}\right)$$

$$= (\cos x)(\cosh y) - (\sin x)\left(-\frac{\sinh y}{i}\right)$$

$$= (\cos x)(\cosh y) - i(\sin x)(\sinh y).$$

Thus

$$|\cos z|^2 = (\cos x)^2(\cosh y)^2 + (\sin x)^2(\sinh y)^2$$

$$= (1 - (\sin x)^2)(\cosh y)^2 + (\sin x)^2 \left(\frac{e^{2y} - 2 + e^{-2y}}{4} \right)$$

$$= (\cosh y)^2 - (\sin x)^2(\cosh y)^2 + (\sin x)^2 \left(\frac{e^{2y} + 2 + e^{-2y}}{4} - 1 \right)$$

$$= (\cosh y)^2 - (\sin x)^2(\cosh y)^2 + (\sin x)^2 \left((\cosh y)^2 - 1 \right)$$

$$= (\cosh y)^2 - \cancel{(\sin x)^2(\cosh y)^2} + \cancel{(\sin x)^2(\cosh y)^2} - (\sin x)^2$$

$$= (\cosh y)^2 - (\sin x)^2.$$

Solution to Exercise 1.38

Let $z = x + iy$, where $x, y \in \mathbb{R}$. Then $\cos z = 3$ gives

$$(\cos x)(\cosh y) = 3, \tag{5.15}$$

$$(\sin x)(\sinh y) = 0. \tag{5.16}$$

We note that $\sinh y = 0$ if and only if $y = 0$. But $y = 0$ is impossible, since this would mean $z \ (= x + iy = x)$ would be real, and there are no real solutions x to $\cos x = 3$! Thus we must have $\sinh y \neq 0$, and so (5.16) implies that $\sin x = 0$. Hence $x \in \{n\pi : n \in \mathbb{Z}\}$. But then $\cos x = \pm 1$. As

$$\cosh y = \frac{e^y + e^{-y}}{2} > 0 \text{ for all } y \in \mathbb{R},$$

we see from (5.15) that $\cos x$ can't be -1. So $x \in \{2\pi n : n \in \mathbb{Z}\}$ and $\cos x = 1$. Then $\cosh y = 3$ yields

$$\frac{e^y + e^{-y}}{2} = 3,$$

that is, $(e^y)^2 - 6e^y + 1 = 0$. Thus

$$e^y = \frac{6 \pm \sqrt{36 - 4}}{2} = 3 \pm \sqrt{9 - 1} = 3 \pm 2\sqrt{2},$$

and so $y = \log(3 + 2\sqrt{2})$ or

$$y = \log(3 - 2\sqrt{2}) = \log \frac{9 - 8}{3 + 2\sqrt{2}} = \log \frac{1}{3 + 2\sqrt{2}} = -\log(3 + 2\sqrt{2}).$$

Consequently, $z \in \{2\pi n \pm i \log(3 + 2\sqrt{2}), n \in \mathbb{Z}\}$.

Vice versa, if $z = 2\pi n \pm i \log(3 + 2\sqrt{2})$ for some $n \in \mathbb{Z}$, then

$$\cos z = \underbrace{(\cos(2\pi n))}_{=1}(\cosh(\pm(3 \pm 2\sqrt{2}))) - i\underbrace{(\sin(2\pi n))}_{=0}(\sinh \cdots))$$

$$= \cosh(\pm(3 \pm 2\sqrt{2})) = \frac{e^{\log(3+2\sqrt{2})} + e^{-\log(3+2\sqrt{2})}}{2}$$

$$= \frac{3 + 2\sqrt{2} + (3 + 2\sqrt{2})^{-1}}{2} = \frac{3 + 2\sqrt{2} + 3 - 2\sqrt{2}}{2} = 3.$$

Consequently, $\cos z = 3$ if and only if $z \in \{2\pi n \pm i \log(3 + 2\sqrt{2}), n \in \mathbb{Z}\}$

Solution to Exercise 1.39

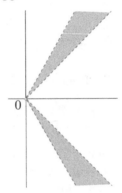

Fig. 5.14   The set $\left\{z \in \mathbb{C} : z \neq 0, \frac{\pi}{4} < |\mathrm{Arg}(z)| < \frac{\pi}{3}\right\}$.

Solution to Exercise 1.40

We have

$$\mathrm{Log}(1 + i) = \mathrm{Log}\left(\sqrt{2}\left(\frac{1}{\sqrt{2}} + i\frac{1}{\sqrt{2}}\right)\right) = \mathrm{Log}\left(\sqrt{2}\left(\cos\frac{\pi}{4} + i\sin\frac{\pi}{4}\right)\right)$$

$$= \mathrm{Log}\left(\sqrt{2}\exp\left(i\frac{\pi}{4}\right)\right) = \log\sqrt{2} + i\frac{\pi}{4}.$$

Solution to Exercise 1.41

We have

$$\mathrm{Log}(-1) = \mathrm{Log}(1 \cdot \exp(i\pi)) = \log 1 + i\pi = 0 + i\pi = i\pi,$$

$$\mathrm{Log}(1) = \mathrm{Log}(1 \cdot \exp(i0)) = \log 1 + i0 = 0 + i0 = 0.$$

With $z = -1$, we have $\text{Log}(z^2) = \text{Log}((-1)^2) = \text{Log}(1) = 0$, while $2 \cdot \text{Log}(z) = 2 \cdot \text{Log}(-1) = 2 \cdot i\pi$. So we see that when $z = -1$,

$$\text{Log}(z^2) = 0 \neq 2 \cdot i\pi = 2 \cdot \text{Log}(z).$$

Solution to Exercise 1.42

Let $\mathbb{A} := \{z \in \mathbb{C} : 1 < z < e\}$. Then $z \in \mathbb{A}$ if and only if $z = r\exp(i\text{Arg}(z))$, where $1 < r < e$ and $\text{Arg}(z) \in (-\pi, \pi]$. For such a z,

$$\text{Log}(z) = \text{Log}(r\exp(i\text{Arg}(z))) = \log r + i\text{Arg}(z)$$

and $0 = \log 1 < \log r < \log e = 1$. So the image lies in the rectangle

$$\mathbb{I} := \{x + iy : 0 < x < 1, \ -\pi < y < \pi\}.$$

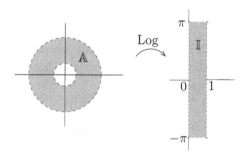

Vice versa, if $x + iy \in \mathbb{I}$, then $z := \exp(x + iy) = e^x \exp(iy) \in \mathbb{A}$ since $|z| = e^x \in (1, e)$, and $\log(z) = \text{Log}(e^x \exp(iy)) = \log e^x + iy = x + iy$. Hence the image of $\mathbb{A}$ under Log is precisely $\mathbb{I}$.

Solution to Exercise 1.43

The principal value of $(1 + i)^{1-i}$ is $\exp((1 - i)\text{Log}(1 + i))$. We have

$$\text{Log}(1 + i) = \text{Log}\left(\sqrt{2}\exp\left(i\frac{\pi}{4}\right)\right) = \log\sqrt{2} + i\frac{\pi}{4}.$$

Thus the principal value of $(1 + i)^{1-i}$ is

$$\begin{aligned}
\exp((1 - i)\text{Log}(1 + i)) &= \exp\left((1 - i)\left(\log\sqrt{2} + i\frac{\pi}{4}\right)\right) \\
&= e^{\log\sqrt{2} + \frac{\pi}{4}}\exp\left(i\left(\frac{\pi}{4} - \log\sqrt{2}\right)\right) \\
&= \sqrt{2}\,e^{\frac{\pi}{4}}\frac{(1 + i)}{\sqrt{2}}\exp(-i\log\sqrt{2}) \\
&= e^{\frac{\pi}{4}}(1 + i)\Big(\cos(\log\sqrt{2}) - i\sin(\log\sqrt{2})\Big).
\end{aligned}$$

Solutions to the exercises from Chapter 2

Solution to Exercise 2.1

For $z \neq 0$, we have

$$\frac{f(z) - f(0)}{z - 0} - 0 = \frac{|z|^2 - 0}{z - 0} = \frac{|z|^2}{z}.$$

Thus given $\epsilon > 0$, we set $\delta = \epsilon > 0$, and then whenever $0 < |z - 0| = |z| < \delta$,

$$\left| \frac{f(z) - f(0)}{z - 0} - 0 \right| = \left| \frac{|z|^2}{z} \right| = \frac{|z|^2}{|z|} = |z| < \delta = \epsilon.$$

Hence f is complex differentiable at 0, and $f'(0) = 0$.

Solution to Exercise 2.2

Let $w_0 \in D^*$. Then $\overline{w_0} \in D$. Since f is holomorphic in D, given $\epsilon > 0$, there exists a $\delta > 0$ such that whenever $0 < |z - \overline{w_0}| < \delta$, $z \in D$ and

$$\left| \frac{f(z) - f(\overline{w_0})}{z - \overline{w_0}} - f'(\overline{w_0}) \right| < \epsilon. \qquad (5.17)$$

Now let w be such that $0 < |w - w_0| < \delta$. But then

$$0 < |w - w_0| = |\overline{w - w_0}| = |\overline{w} - \overline{w_0}| < \delta,$$

and so $\overline{w} \in D$. Thus $w \in D^*$. Moreover,

$$
\left| \frac{f^*(w) - f^*(w_0)}{w - w_0} - \overline{f'(\overline{w_0})} \right| = \left| \frac{\overline{f(\overline{w})} - \overline{f(\overline{w_0})}}{w - w_0} - \overline{f'(\overline{w_0})} \right|
$$

$$
= \left| \overline{\frac{f(\overline{w}) - f(\overline{w_0})}{\overline{w} - \overline{w_0}}} - \overline{f'(\overline{w_0})} \right|
$$

$$
= \left| \frac{f(\overline{w}) - f(\overline{w_0})}{\overline{w} - \overline{w_0}} - f'(\overline{w_0}) \right| < \epsilon \quad \text{(using (5.17))}.
$$

So f^* is complex differentiable at w_0 and $(f^*)'(w_0) = \overline{f'(\overline{w_0})}$. As $w_0 \in D^*$ was arbitrary, f^* is holomorphic in D^*.

Solution to Exercise 2.3

Since f is complex differentiable at z_0, there exists an $r > 0$ and a function $h : D(z_0, r) \to \mathbb{C}$, where $D(z_0, r) := \{ z \in \mathbb{C} : |z - z_0| < r \} \subset D$ such that

$$f(z) = f(z_0) + (f'(z_0) + h(z))(z - z_0) \text{ for } |z - z_0| < r,$$

and

$$\lim_{z \to z_0} h(z) = 0.$$

We can choose an $r' < r$ so that in $D_*(z_0, r') := \{z \in \mathbb{C} : 0 < |z - z_0| < r'\}$ ($\subset D(z_0, r) \subset D$), we have $|h(z)| < 1$. Now given $\epsilon > 0$, set

$$\delta = \min \left\{ \frac{\epsilon}{|f'(z_0)| + 1}, r' \right\}.$$

Then whenever $0 < |z - z_0| < \delta$, we have $z \in D(z_0, r')$ and so

$$|f(z) - f(z_0)| = |f'(z_0) + h(z)||z - z_0| \leq (|f'(z_0)| + |h(z)|)\frac{\epsilon}{|f'(z_0)| + 1}$$

$$< (|f'(z_0)| + 1)\frac{\epsilon}{|f'(z_0)| + 1} = \epsilon.$$

Hence f is continuous at z_0.

Solution to Exercise 2.4

Using the fact that $f, g : U \to \mathbb{C}$ are complex differentiable functions at $z_0 \in U$, it follows from Lemma 2.1 that there exists an $r > 0$, and functions $h_f, h_g : D(z_0, r) \to \mathbb{C}$, where $D(z_0, r) := \{z \in \mathbb{C} : |z - z_0| < r\}$, such that for $|z - z_0| < r$

$$f(z) = f(z_0) + (f'(z_0) + h_f(z))(z - z_0), \tag{5.18}$$

$$g(z) = g(z_0) + (g'(z_0) + h_g(z))(z - z_0), \tag{5.19}$$

and $\lim_{z \to z_0} h_f(z) = 0 = \lim_{z \to z_0} h_g(z)$.

(1) Adding (5.18) and (5.19), we obtain for $|z - z_0| < r$ that

$$(f + g)(z) = (f + g)(z_0) + \left(f'(z_0) + g'(z_0) + h_{f+g}(z)\right)(z - z_0),$$

where $h_{f+g}(z) := h_f(z) + h_g(z)$ in $D(z_0, r)$. Moreover,

$$\lim_{z \to z_0} h_{f+g}(z) = \lim_{z \to z_0}(h_f(z) + h_g(z)) = \lim_{z \to z_0} h_f(z) + \lim_{z \to z_0} h_g(z) = 0 + 0 = 0.$$

So by Lemma 2.1, it follows that $f + g$ is complex differentiable and $(f + g)'(z_0) = f'(z_0) + g'(z_0)$.

(2) Multiplying (5.18) by α, we obtain for $|z - z_0| < r$ that

$$(\alpha \cdot f)(z) = (\alpha \cdot f)(z_0) + \left(\alpha \cdot f'(z_0) + h_{\alpha \cdot f}(z)\right)(z - z_0),$$

where $h_{\alpha \cdot f}(z) := \alpha \cdot h_f(z)$ in $D(z_0, r)$. Moreover,

$$\lim_{z \to z_0} h_{\alpha \cdot f}(z) = \lim_{z \to z_0}(\alpha \cdot h_f(z)) = \alpha \cdot \lim_{z \to z_0} h_f(z) = \alpha \cdot 0 = 0.$$

So by Lemma 2.1, it follows that $\alpha \cdot f$ is complex differentiable and $(\alpha \cdot f)'(z_0) = \alpha \cdot f'(z_0)$.

(3) Multiplying (5.18) and (5.19), we obtain for $|z - z_0| < r$ that

$$(fg)(z) = (fg)(z_0) + \Big(f'(z_0)g(z_0) + f(z_0)g'(z_0) + h_{fg}(z)\Big)(z - z_0),$$

where

$$h_{fg}(z) := f(z_0)h_g(z) + g(z_0)h_f(z) + (z - z_0)(f'(z_0) + h_f(z))(g'(z_0) + h_g(z))$$

in $D(z_0, r)$. Moreover,

$$\lim_{z \to z_0} h_{fg}(z) = f(z_0) \cdot 0 + g(z_0) \cdot 0 + 0 \cdot (f'(z_0) + 0) \cdot (g'(z_0) + 0) = 0.$$

So by Lemma 2.1, it follows that fg is complex differentiable and

$$(fg)'(z_0) = f'(z_0)g(z_0) + f(z_0)g'(z_0).$$

Solution to Exercise 2.5

Suppose that $\mathrm{Hol}(\mathbb{D})$ is finite dimensional with dimension d. Then the $d+1$ vectors $1, z, z^2, \cdots, z^d \in \mathrm{Hol}(\mathbb{D})$ must be linearly dependent. So there exist scalars $\alpha_0, \cdots, \alpha_d$, not all zeros, such that

$$\alpha_0 \cdot 1 + \alpha_1 \cdot z + \cdots + \alpha_d \cdot z^d = 0 \quad (z \in \mathbb{D}).$$

Let $k \in \{0, 1, \cdots, d\}$ be the smallest index such that $\alpha_k \neq 0$. Then by differentiating k times and evaluating at $0 \in \mathbb{D}$, we get

$$0 + \alpha_k \cdot k! + 0 = 0,$$

and so $\alpha_k = 0$, a contradiction.

Solution to Exercise 2.6

Let $z_0 \in U$. Since f is holomorphic at z_0, there exists a $r > 0$ and a complex-valued h defined on $D(z_0, r) := \{z \in \mathbb{C} : |z - z_0| < r\} \subset U$ such that

$$f(z) = f(z_0) + (f'(z_0) + h(z))(z - z_0), \quad z \in D(z_0, r),$$

and

$$\lim_{z \to z_0} h(z) = 0. \tag{5.20}$$

Thus with $g := 1/f$, we have

$$\frac{1}{g(z)} = \frac{1}{g(z_0)} + (f'(z_0) + h(z))(z - z_0),$$

and so $g(z_0) = g(z) + (f'(z_0) + h(z))g(z_0)g(z) \cdot (z - z_0)$. Rearranging, we obtain

$$g(z) = g(z_0) + (-f'(z_0)g(z_0)g(z) - h(z)g(z_0)g(z)) \cdot (z - z_0)$$

$$= g(z_0) + \left(-\frac{f'(z_0)}{(f(z_0))^2} + \frac{f'(z_0)}{(f(z_0))^2} - \frac{f'(z_0)}{f(z_0)f(z)} - \frac{h(z)}{f(z_0)f(z)} \right)(z - z_0)$$

$$= g(z_0) + \left(-\frac{f'(z_0)}{(f(z_0))^2} + \varphi(z) \right) \cdot (z - z_0),$$

where

$$\varphi(z) := \frac{f'(z_0)}{(f(z_0))^2} - \frac{f'(z_0)}{f(z_0)f(z)} - \frac{h(z)}{f(z_0)f(z)}, \quad z \in D(z_0, r).$$

Using the continuity of f at z_0 and (5.20), we obtain

$$\lim_{z \to z_0} \varphi(z) = \frac{f'(z_0)}{(f(z_0))^2} - \frac{f'(z_0)}{f(z_0)f(z_0)} - \frac{0}{f(z_0)f(z)} = 0.$$

Hence g is differentiable at z_0 and

$$g'(z_0) = -\frac{f'(z_0)}{(f(z_0))^2}.$$

Solution to Exercise 2.7

We have already shown this for $m \geq 0$. Suppose now that $m = -n$ for some $n \in \mathbb{N}$. Then consider the map

$$z \mapsto z^m = z^{-n} = \frac{1}{z^n} = \frac{1}{f(z)},$$

where $f(z) := z^n$, $z \in \mathbb{C} \setminus \{0\}$. Since f is holomorphic and pointwise nonzero in $\mathbb{C} \setminus \{0\}$, it follows that $1/f$ is holomorphic too, and

$$\left(\frac{1}{f} \right)'(z) = -\frac{f'(z)}{(f(z))^2} = -\frac{nz^{n-1}}{(z^n)^2} = -n \cdot \frac{1}{z^{n+1}} = m \cdot \frac{1}{z^{-m+1}} = mz^{m-1}.$$

This completes the proof.

Solution to Exercise 2.8

Let $f : \mathbb{D} \to \mathbb{C}$ be defined by

$$f(z) = -\frac{1+z}{1-z}, \quad z \in \mathbb{D},$$

and $g : \mathbb{C} \to \mathbb{C}$ be defined by $g(z) = \exp z$, $z \in \mathbb{C}$. Then $f(\mathbb{D}) \subset \mathbb{C} = D_g$, and so $g \circ f$ is holomorphic in $\mathbb{D}$, and moreover,

$$
\begin{aligned}
(g \circ f)'(z) &= g'(f(z)) \cdot f'(z) = \exp\left(-\frac{1+z}{1-z}\right) \cdot \frac{d}{dz}\left(-\frac{1+z}{1-z}\right) \\
&= \exp\left(-\frac{1+z}{1-z}\right) \cdot \left(-(1+z)\frac{d}{dz}\left(\frac{1}{1-z}\right) - \frac{1}{1-z}\frac{d}{dz}(1+z)\right) \\
&= \exp\left(-\frac{1+z}{1-z}\right) \cdot \left(-\frac{(1+z)}{(1-z)^2} - \frac{1}{1-z}\right) \\
&= -\frac{2}{(1-z)^2}\exp\left(-\frac{1+z}{1-z}\right).
\end{aligned}
$$

Thus $\dfrac{d}{dz}\left(\exp\left(-\dfrac{1+z}{1-z}\right)\right) = -\dfrac{2}{(1-z)^2}\exp\left(-\dfrac{1+z}{1-z}\right)$, $z \in \mathbb{D}$.

Solution to Exercise 2.9

Let $z = x + iy$, where $x, y \in \mathbb{R}$. We have $|z|^2 = x^2 + y^2$. So if u, v denote the real part and the imaginary part of $|z|^2$, then we have $u = x^2 + y^2$, $v = 0$. So

$$
\frac{\partial u}{\partial x} = 2x, \quad \frac{\partial v}{\partial y} = 0,
$$

$$
\frac{\partial u}{\partial y} = 2y, \quad \frac{\partial v}{\partial x} = 0.
$$

Since $z \neq 0$, it follows that x or y is nonzero. So at least one of the Cauchy-Riemann equations is violated, since either

$$
\left(\frac{\partial u}{\partial x} = \right) 2x \neq 0 \left(= \frac{\partial v}{\partial y}\right), \text{ or}
$$

$$
\left(\frac{\partial u}{\partial y} = \right) 2y \neq 0 \left(= -\frac{\partial v}{\partial y}\right).
$$

So $|z|^2$ is not differentiable at any nonzero complex number.

Solution to Exercise 2.10

Let $z = x + iy$, where $x, y \in \mathbb{R}$. Then

$$
\begin{aligned}
z^3 &= (x + iy)^3 = x^3 + 3x^2(iy) + 3x(iy)^2 + (iy)^3 \\
&= x^3 - 3xy^2 + i(3x^2y - y^3).
\end{aligned}
$$

So if u, v denote the real and imaginary parts of z^3, then we have

$$u(x, y) = x^3 - 3xy^2 \text{ and}$$
$$v(x, y) = 3x^2 y - y^3.$$

We have u, v are continuously differentiable, and

$$\frac{\partial u}{\partial x} = 3x^2 - 3y^2 = \frac{\partial v}{\partial y}, \text{ and}$$
$$\frac{\partial u}{\partial y} = -6xy = -\frac{\partial v}{\partial x}.$$

So the Cauchy-Riemann equations are satisfied everywhere in $\mathbb{R}^2$. Hence $z \mapsto z^3$ is entire.

Solution to Exercise 2.11

Let $z = x + iy$, where $x, y \in \mathbb{R}$. We have $\mathrm{Re}(z) = \mathrm{Re}(x + iy) = x$. So if u, v denote the real and imaginary parts of $\mathrm{Re}(z)$, then we have

$$u = x,$$
$$v = 0.$$

So we have

$$\frac{\partial u}{\partial x} = 1 \neq 0 = \frac{\partial v}{\partial y} \text{ for all } (x, y) \in \mathbb{R}^2.$$

Thus the Cauchy-Riemann equations are satisfied nowhere in $\mathbb{R}^2$. Hence $\mathrm{Re}(z)$ is not complex differentiable at any point in $\mathbb{C}$.

Solution to Exercise 2.12

Let u, v be the real and imaginary parts of f. Then $v = 0$. Thus

$$\frac{\partial u}{\partial x} = \frac{\partial v}{\partial y} = 0 \text{ and } \frac{\partial u}{\partial y} = -\frac{\partial v}{\partial x} = 0.$$

Hence

$$u(x, y_0) - u(x_0, y_0) = \int_{x_0}^{x} \frac{\partial u}{\partial x}(\xi, y_0) d\xi = 0$$

whenever the straight line segment joining $(x, y_0) \in D$ to $(x_0, y_0) \in D$ lies in D. Similarly,

$$u(x_0, y) - u(x_0, y_0) = \int_{y_0}^{y} \frac{\partial u}{\partial y}(x_0, \eta) d\eta = 0$$

whenever the straight line segment joining $(x_0, y_0) \in D$ to $(x_0, y) \in D$ lies in D. So the value of u along horizontal and vertical line segments lying inside D is constant. As D is path-connected, it now follows that u is constant in D (because any two points in D can be joined by a stepwise path). Hence $f = u + i0 = u$ is constant in D.

Solution to Exercise 2.13

Let u, v be the real and imaginary parts of f. Then $f'(z) = \dfrac{\partial u}{\partial x} + i \dfrac{\partial v}{\partial x} = 0$ gives

$$\frac{\partial u}{\partial x} = \frac{\partial v}{\partial x} = 0 \text{ in } D,$$

and using the Cauchy-Riemann equations, also that

$$\frac{\partial u}{\partial y} \left(= -\frac{\partial v}{\partial x} \right) = 0 \text{ and } \frac{\partial v}{\partial y} \left(= \frac{\partial u}{\partial x} \right) = 0.$$

Hence

$$u(x, y_0) - u(x_0, y_0) = \int_{x_0}^{x} \frac{\partial u}{\partial x} (\xi, y_0) d\xi = 0$$

whenever the straight line segment joining $(x, y_0) \in D$ to $(x_0, y_0) \in D$ lies in D. Similarly,

$$u(x_0, y) - u(x_0, y_0) = \int_{y_0}^{y} \frac{\partial u}{\partial y} (x_0, \eta) d\eta = 0$$

whenever the straight line segment joining $(x_0, y_0) \in D$ to $(x_0, y) \in D$ lies in D. So the value of u along horizontal and vertical line segments lying inside D is constant. As D is path-connected, it now follows that u is constant in D (because any two points in D can be joined by a stepwise path). Similarly v is also constant in D. Consequently, $f = u + iv$ is constant in D.

Solution to Exercise 2.14

We have using the chain rule that

$$\frac{\partial u}{\partial x}(x, y) = h'(v(x, y)) \frac{\partial v}{\partial x}(x, y), \qquad \frac{\partial u}{\partial y}(x, y) = h'(v(x, y)) \frac{\partial v}{\partial y}(x, y).$$

Using these and the Cauch-Riemann equations, we obtain

$$\frac{\partial u}{\partial y}(x,y) = h'(v(x,y)) \cdot \frac{\partial v}{\partial y}(x,y) = h'(v(x,y)) \cdot \frac{\partial u}{\partial x}(x,y)$$

$$= h'(v(x,y)) \cdot \left(h'(v(x,y)) \frac{\partial v}{\partial x}(x,y) \right) = (h'(v(x,y)))^2 \cdot \frac{\partial v}{\partial x}(x,y)$$

$$= -(h'(v(x,y)))^2 \cdot \frac{\partial u}{\partial y}(x,y),$$

and so $(1 + (h'(v(x,y)))^2) \frac{\partial u}{\partial y}(x,y) = 0$. As $(1 + (h'(v(x,y)))^2) \geq 1 > 0$,

$$\frac{\partial u}{\partial y}(x,y) = 0.$$

Using the Cauchy-Riemann equations, we also obtain

$$\frac{\partial v}{\partial x}(x,y) = -\frac{\partial u}{\partial y}(x,y) = 0.$$

From this, it follows that

$$\frac{\partial u}{\partial x}(x,y) = h'(v(x,y)) \frac{\partial v}{\partial x}(x,y) = h'(v(x,y)) \cdot 0 = 0,$$

and using the Cauchy-Riemann equations again, we have

$$\frac{\partial v}{\partial y}(x,y) = \frac{\partial u}{\partial x}(x,y) = 0.$$

But now it follows that u is constant locally along horizontal and along vertical lines. Since D is a domain, it is path-connected, and so we know that any two points in D can be connected by a stepwise path. This implies that u is constant in D. Similarly v is constant in D. So $f = u + iv$ is constant in D as well.

Solution to Exercise 2.15

("If" part) Suppose $k = 2$. Then

$$f(z) = x^2 - y^2 + 2xyi = x^2 + (iy)^2 + 2x(iy) = (x + iy)^2 = z^2,$$

which we have seen is entire in Example 2.1.

("Only if" part) Now suppose that f is entire. Then the Cauchy-Riemann equations must be satisfied everywhere in $\mathbb{R}^2$. So in particular,

$$\frac{\partial u}{\partial x} = 2x = kx = \frac{\partial v}{\partial y} \text{ for all } x, y \in \mathbb{R}.$$

If we take $x = 1$, then the above yields that $k = 2$.

Solution to Exercise 2.16

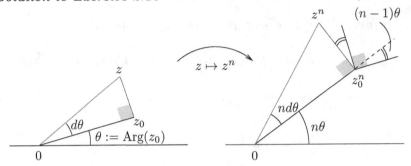

We have that the length of $z - z_0$ is $|z_0| \tan(d\theta) \approx |z_0| d\theta$, while the length of $z^n - z_0^n$ is $|z_0|^n \tan(nd\theta) \approx |z_0|^n nd\theta$, and so the magnification produced locally by $z \mapsto z^n$ is given by

$$\frac{|z^n - z_0^n|}{|z - z_0|} \approx \frac{|z_0|^n nd\theta}{|z_0| d\theta} = n|z_0|^{n-1}.$$

Also, from the picture, we see that the anticlockwise rotation produced locally is equal to $(n-1)\theta$. Hence

$$f'(z_0) = n|z_0|^{n-1}(\cos((n-1)\theta) + i\sin((n-1)\theta))$$
$$= n(|z_0|(\cos\theta + i\sin\theta))^{n-1} = nz_0^{n-1}.$$

Consequently, $\dfrac{d}{dz}z^n = nz^{n-1}$, $z \in \mathbb{C}$.

Solution to Exercise 2.17

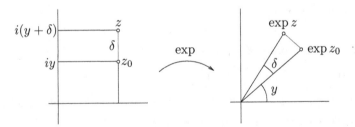

Fig. 5.15   Calculation of the amount of local magnification produced by exp.

From Figure 5.15, we see that the magnification produced is

$$\frac{e^x \cdot \delta}{\delta} = e^x.$$

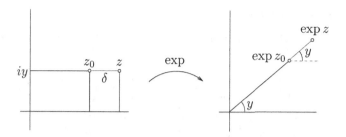

From the picture, we see that the anticlockwise rotation produced is y. Thus the complex derivative at z_0 must be $e^x(\cos y + i \sin y) = \exp(x + iy)$. Consequently, $\exp' z = \exp z$.

Solution to Exercise 2.18

Let $z_0 \in \mathbb{C}$. Move the point z_0 along the line with slope 1 through a distance of δ to get the point z. Similarly move z_0 horizontally to the left by a distance δ to get the point $\tilde{z}$. Suppose that the real part mapping $\text{Re}(\cdot)$ is complex differentiable at z_0. In Figure 5.16, by looking at the images of $z, \tilde{z}$ under the mapping $\text{Re}(\cdot)$, we see get conflicting values of the local rotation produced as being $45°$ and $0°$, which cannot happen. So the map cannot be complex differentiable at z_0. As the choice of $z_0 \in \mathbb{C}$ was arbitrary, the map is complex differentiable nowhere.

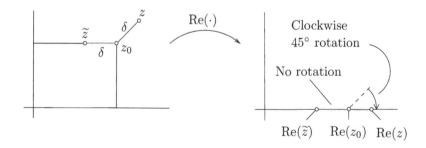

Fig. 5.16 Non complex differentiability of $\text{Re}(\cdot)$.

Solution to Exercise 2.19

For $f = u + iv$ with smooth $u, v \in C^2$, that is, twice continuously differentiable, we have

$$4\frac{\partial}{\partial z}\frac{\partial}{\partial \bar{z}}f = 4 \cdot \frac{1}{2}\left(\frac{\partial}{\partial x} - i\frac{\partial}{\partial y}\right) \cdot \frac{1}{2}\left(\frac{\partial u}{\partial x} + i\frac{\partial u}{\partial y} + i\left(\frac{\partial v}{\partial x} + i\frac{\partial v}{\partial y}\right)\right)$$

$$= \left(\frac{\partial}{\partial x} - i\frac{\partial}{\partial y}\right)\left(\frac{\partial u}{\partial x} + i\frac{\partial u}{\partial y} + i\left(\frac{\partial v}{\partial x} + i\frac{\partial v}{\partial y}\right)\right)$$

$$= \frac{\partial^2 u}{\partial x^2} - \frac{\partial^2 v}{\partial x \partial y} + i\frac{\partial^2 u}{\partial x \partial y} + i\frac{\partial^2 v}{\partial x^2} - i\frac{\partial^2 u}{\partial y \partial x} + i\frac{\partial^2 v}{\partial y^2} + \frac{\partial^2 u}{\partial y^2} + \frac{\partial^2 v}{\partial y \partial x}$$

$$= \frac{\partial^2 u}{\partial x^2} + \frac{\partial^2 u}{\partial y^2} + i\left(\frac{\partial^2 v}{\partial x^2} + \frac{\partial^2 v}{\partial y^2}\right) + i\left(\frac{\partial^2 u}{\partial x \partial y} - \frac{\partial^2 u}{\partial y \partial x}\right) + \frac{\partial^2 v}{\partial y \partial x} - \frac{\partial^2 v}{\partial x \partial y}.$$

But since $u, v \in C^2$, we have $\dfrac{\partial^2 u}{\partial x \partial y} - \dfrac{\partial^2 u}{\partial y \partial x} = 0 = \dfrac{\partial^2 v}{\partial y \partial x} - \dfrac{\partial^2 v}{\partial x \partial y}$. Thus

$$4\frac{\partial}{\partial z}\frac{\partial}{\partial \bar{z}}f = \frac{\partial^2 u}{\partial x^2} + \frac{\partial^2 u}{\partial y^2} + i\left(\frac{\partial^2 v}{\partial x^2} + \frac{\partial^2 v}{\partial y^2}\right) = \left(\frac{\partial^2}{\partial x^2} + \frac{\partial^2}{\partial y^2}\right)(u + iv) = \Delta f.$$

Solutions to the exercises from Chapter 3

Solution to Exercise 3.1

We have $\gamma_1 = \cos t + i \sin t$, $\gamma_2 = \cos(2t) + i \sin(2t)$, and $\gamma_3 = \cos t - i \sin t$, and in each case, $(\mathrm{Re}(\gamma_k(t)))^2 + (\mathrm{Im}(\gamma_k))^2 = 1$, $k = 1, 2, 3$, and so the image of γ_k is contained in the circle $\mathbb{T}$ with center 0 and radius 1. Also if $z = \exp(i\theta)$ with $\theta \in [0, 2\pi)$, then $z = \gamma_1(\theta) = \gamma_2(\theta/2) = \gamma_3(2\pi - \theta)$, and so every point on $\mathbb{T}$ belongs to the image of each of the curves $\gamma_1, \gamma_2, \gamma_3$. We have

$$\int_{\gamma_1} \frac{1}{z} dz = \int_0^{2\pi} \frac{1}{\exp(it)} \cdot i \exp(it) dt = 2\pi i,$$

$$\int_{\gamma_2} \frac{1}{z} dz = \int_0^{2\pi} \frac{1}{\exp(2it)} \cdot 2i \exp(2it) dt = 4\pi i,$$

$$\int_{\gamma_3} \frac{1}{z} dz = \int_0^{2\pi} \frac{1}{\exp(-it)} \cdot (-i) \exp(-it) dt = -2\pi i.$$

Solution to Exercise 3.2

Let $\gamma(t) = x(t) + iy(t)$, $t \in [0, 1]$, where x, y are real-valued. Also, let u, v be the real and imaginary parts of f, respectively. Then

$$f'(\gamma(t)) \cdot \gamma'(t) = \left(\frac{\partial u}{\partial x}(x(t), y(t)) + i \frac{\partial v}{\partial x}(x(t), y(t)) \right) (x'(t) + iy'(t))$$

$$= \frac{\partial u}{\partial x}(x(t), y(t)) \cdot x'(t) - \frac{\partial v}{\partial x}(x(t), y(t)) \cdot y'(t)$$

$$+ i \left(\frac{\partial u}{\partial x}(x(t), y(t)) \cdot y'(t) + \frac{\partial v}{\partial x}(x(t), y(t)) \cdot x'(t) \right)$$

$$= \frac{\partial u}{\partial x}(x(t), y(t)) \cdot x'(t) + \frac{\partial u}{\partial y}(x(t), y(t)) \cdot y'(t)$$

$$+ i \left(\frac{\partial v}{\partial y}(x(t), y(t)) \cdot y'(t) + \frac{\partial v}{\partial x}(x(t), y(t)) \cdot x'(t) \right)$$

$$\text{(using Cauchy-Riemann equations)}$$

$$= \frac{d}{dt} u(x(t), y(t)) + i \frac{d}{dt} v(x(t), y(t)) \quad \text{(using the Chain Rule)}$$

$$= \frac{d}{dt} (u(x(t), y(t)) + iv(x(t), y(t))) = \frac{d}{dt} f(\gamma(t)).$$

Solution to Exercise 3.3

Let γ be the circular path $\gamma(t) = 2\exp(it)$, $t \in [0, 2\pi]$.

(1) We have

$$\int_\gamma (z + \bar{z})dz = \int_0^{2\pi} (2\exp(it) + 2\exp(-it)) \cdot 2i \cdot \exp(it)dt$$

$$= 4i \int_0^{2\pi} (\exp(2it) + 1)dt = 4i \cdot 0 + 4i \cdot 2\pi = 8\pi i.$$

(2) We have

$$\int_\gamma (z^2 - 2z + 3)dz = \int_0^{2\pi} (4\exp(2it) - 4\exp(it) + 3) \cdot 2i \cdot \exp(it)dt$$

$$= \int_0^{2\pi} i(8\exp(3it) - 8\exp(2it) + 6\exp(it))dt$$

$$= 0 + 0 + 0 = 0.$$

(3) We have

$$\int_\gamma xy\,dz = \int_0^{2\pi} 2\cos t \cdot 2\sin t \cdot 2i \cdot (\cos t + i\sin t)dt$$

$$= 4i \int_0^{2\pi} (\sin(2t))(\cos t + i\sin t)dt$$

$$= 4i \int_{-\pi}^{\pi} \underbrace{(\sin(2t))\cos t}_{\text{odd function}} dt - 2\int_0^{2\pi} (\cos t - \cos(3t))dt$$

$$= 0 - 2(0 - 0) = 0.$$

Solution to Exercise 3.4

(1) $\gamma(t) = (1 + i)t$, $t \in [0, 1]$, and so $\displaystyle\int_\gamma \operatorname{Re}(z)dz = \int_0^1 t(1 + i)dt = \dfrac{1 + i}{2}$.

(2) $\gamma(t) = i + \exp(it)$, $t \in [-\pi/2, 0]$, and so

$$\int_\gamma \operatorname{Re}(z)dz = \int_{-\pi/2}^0 (\cos t)i\exp(it)dt = \int_{-\pi/2}^0 i(\cos t)^2 - (\cos t)(\sin t)dt$$

$$= \int_{-\pi/2}^0 \left(i\frac{\cos(2t) + 1}{2} - \frac{\sin(2t)}{2}\right)dt$$

$$= 0 + i \cdot \frac{1}{2} \cdot \frac{\pi}{2} + \frac{1}{2} = \frac{1}{2} + i\frac{\pi}{4}.$$

(3) $\gamma(t) = t + it^2$, $t \in [0, 1]$, and so

$$\int_\gamma \operatorname{Re}(z)dz = \int_0^1 t \cdot (1 + 2it)dt = \frac{1}{2} + 2i \cdot \frac{1}{3} = \frac{1}{2} + \frac{2}{3}i.$$

Solution to Exercise 3.5

By the Binomial Theorem,

$$(1 + z)^n = \sum_{\ell=0}^n \binom{n}{\ell} z^\ell 1^{n-\ell} = \sum_{\ell=0}^n \binom{n}{\ell} z^\ell.$$

Thus for $0 \le k \le n$,

$$\frac{(1 + z)^n}{z^{k+1}} = \sum_{\ell=0}^n \binom{n}{\ell} z^{\ell-k-1},$$

and so

$$\frac{1}{2\pi i} \int_C \frac{(1 + z)^n}{z^{k+1}}dz = \frac{1}{2\pi i} \int_C \sum_{\ell=0}^n \binom{n}{\ell} z^{\ell-k-1}dz$$

$$= \sum_{\ell=0}^n \binom{n}{\ell} \frac{1}{2\pi i} \int_C z^{\ell-k-1}dz = \binom{n}{k}.$$

Solution to Exercise 3.6

Let $U_f, V_f, U_g, V_g : [a, b] \to \mathbb{R}$ be such that $f(\gamma(t))\gamma'(t) = U_f(t) + iV_f(t)$, $g(\gamma(t))\gamma'(t) = U_g(t) + iV_g(t)$, $t \in [a, b]$. Then

$$\int_\gamma (f + g)(z)dz = \int_a^b (f + g)(\gamma(t)) \cdot \gamma'(t)dt$$

$$= \int_a^b \left(f(\gamma(t)) \cdot \gamma'(t) + g(\gamma(t)) \cdot \gamma'(t) \right)dt$$

$$= \int_a^b (U_f(t) + U_g(t))dt + i \int_a^b (V_f(t) + V_g(t))dt$$

$$= \int_a^b U_f(t)dt + \int_a^b U_g(t)dt + i \int_a^b V_f(t)dt + \int_a^b V_g(t)dt$$

$$= \int_\gamma f(z)dz + \int_\gamma g(z)dz.$$

(2) Let $\alpha = p + iq$ where $p, q \in \mathbb{R}$, and let $U, V : [a, b] \to \mathbb{R}$ be such that $f(\gamma(t)) \cdot \gamma'(t) = U(t) + iV(t)$, $t \in [a, b]$. Then

$$\int_\gamma (\alpha \cdot f)(z)dz = \int_a^b (p + iq)(U(t) + iV(t))dt$$

$$= \int_a^b (pU(t) - qV(t))dt + i\int_a^b (pV(t) + qU(t))dt$$

$$= p\left(\int_a^b U(t)dt + i\int_a^b V(t)dt\right) + iq\left(\int_a^b U(t)dt + i\int_a^b V(t)dt\right)$$

$$= (p + iq)\left(\int_a^b U(t)dt + i\int_a^b V(t)dt\right) = \alpha \cdot \int_\gamma f(z)dz.$$

Solution to Exercise 3.7

$-(-\gamma) : [a, b] \to \mathbb{C}$ is given by

$$(-(-\gamma))(t) = (-\gamma)(a + b - t) = \gamma(a + b - (a + b - t)) = \gamma(t), \quad t \in [a, b],$$

and so $-(-\gamma) = \gamma$. (This is obvious visually.)

Solution to Exercise 3.8

We do have that $\gamma(b) = (-\gamma)(a)$, and so the two paths γ and $-\gamma$ can be concatenated, and we have

$$\int_{\gamma + (-\gamma)} f(z)dz = \int_\gamma f(z)dz + \int_{-\gamma} f(z)dz = \int_\gamma f(z)dz - \int_\gamma f(z)dz = 0.$$

Solution to Exercise 3.9

Let $\gamma : [0, 1] \to \mathbb{C}$ be defined by $\gamma(t) = (1 + i)t$, $t \in [0, 1]$. By Pythagoras's Theorem, the length of γ is $\sqrt{1^2 + 1^2} = \sqrt{2}$.

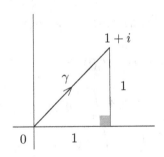

Also, $|(\gamma(t))^2| = |t + it|^2 = 2t^2$, and so $\max\limits_{t\in[0,1]} |(\gamma(t))^2| = 2 \cdot 1^2 = 2$. Thus

$$\left| \int_\gamma z^2 dz \right| \le \left(\max\limits_{t\in[0,1]} |(\gamma(t))^2| \right) \cdot (\text{length of } \gamma) = 2\sqrt{2}.$$

We have

$$\int_\gamma z^2 dz = \int_0^1 (t + it)^2 \cdot (1 + i) dt = \int_0^1 (1 + i)^3 t^2 dt = \frac{(1 + i)^3}{3},$$

and so $\left| \int_\gamma z^2 dz \right| = \dfrac{(\sqrt{2})^3}{3} = \dfrac{2\sqrt{2}}{3}$.

Solution to Exercise 3.10

We have

$$\begin{aligned}
\binom{2n}{n} = \left| \binom{2n}{n} \right| &= \left| \frac{1}{2\pi i} \int_C \frac{(1 + z)^{2n}}{z^{n+1}} dz \right| \\
&\le \frac{1}{2\pi} \left(\max\limits_{|z|=1} \left| \frac{(1 + z)^{2n}}{z^{n+1}} \right| \right) \cdot 2\pi \cdot 1 = \max\limits_{|z|=1} \frac{|1 + z|^{2n}}{1} \\
&\le (1 + 1)^{2n} = 2^{2n} = 4^n.
\end{aligned}$$

Solution to Exercise 3.11

Suppose that $F = U + iV$ is a primitive of $\bar{z}$ in $\mathbb{C}$. Then

$$\frac{\partial U}{\partial x} + i\frac{\partial V}{\partial x} = \frac{\partial V}{\partial y} - i\frac{\partial U}{\partial y} = F' = \bar{z} = x - iy \text{ in } \mathbb{R}^2.$$

Fix $x_0 \in \mathbb{R}$. Then for $(x, y) \in \mathbb{R}^2$, we have

$$V(x, y) - V(x_0, y) = \int_{x_0}^x \frac{\partial V}{\partial x}(\xi, y) d\xi = \int_{x_0}^x -y\, d\xi = -xy + x_0 y.$$

So $V(x, y) = -xy + \varphi(y)$, where $\varphi(y) := V(x_0, y) + x_0 y$. Hence

$$x = \frac{\partial V}{\partial y} = -x + \varphi'(y),$$

that is, $\varphi'(y) = 2x$ for all $x \in \mathbb{R}$, which is clearly impossible, since in particular, we would obtain $2 \cdot 1 = 2 = \varphi'(y) = 2 \cdot 0 = 0$!

Solution to Exercise 3.12

The map $\zeta \mapsto f(\zeta)g'(\zeta) + f'(\zeta)g(\zeta)$ has a primitive since $(fg)' = fg' + f'g$. So by the Fundamental Theorem of Contour Integration,

$$\int_\gamma \Big(f(\zeta)g'(\zeta) + f'(\zeta)g(\zeta)\Big)d\zeta = f(z)g(z) - f(w)g(w),$$

and a rearrangement proves the claim.

Solution to Exercise 3.13

We have $\sin' z = \cos z$, and so $\cos z$ has a primitive in $\mathbb{C}$. Thus by the Fundamental Theorem of Contour Integration

$$\int_\gamma \cos z \, dz = \sin i - \sin(-i) = 2\sin i = 2\frac{\exp(i \cdot i) - \exp(-i \cdot i)}{2i} = \frac{e^{-1} - e^1}{i}$$

$$= \left(e - \frac{1}{e}\right)i.$$

Solution to Exercise 3.14

Since $\exp' z = \exp z$ in $\mathbb{C}$, we have

$$\int_\gamma \exp z \, dz = \exp(a+ib) - \exp 0 = e^a(\cos b + i \sin b) - 1 = e^a \cos b - 1 + ie^a \sin b,$$

for a path γ joining 0 to $a + ib$. If $\gamma(x) = (a + ib)x$, $x \in [0,1]$, then

$$\int_\gamma \exp z \, dz = \int_0^1 \exp(a+ib) \cdot (a+ib)dx = \int_0^1 e^{ax}(\cos(bx) + i\sin(bx))(a+ib)dx.$$

Hence $(a - ib)\displaystyle\int_\gamma \exp z \, dz = \int_0^1 e^{ax}(\cos(bx) + i\sin(bx))(a^2 + b^2)dx$. Thus

$$(a^2 + b^2)\int_0^1 e^{ax}\cos(bx)dx = \mathrm{Re}\left((a - ib)\int_\gamma \exp z \, dz\right)$$

$$= \mathrm{Re}((a - ib)(e^a \cos b - 1 + ie^a \sin b))$$

$$= a(e^a \cos b - 1) + be^a \sin b.$$

Hence $\displaystyle\int_0^1 e^{ax}\cos(bx)dx = \frac{a(e^a \cos b - 1) + be^a \sin b}{a^2 + b^2}.$

Solution to Exercise 3.15

Consider the closed circular path C with center at 0 and radius $r > 0$ traversed in the anticlockwise direction: $C(\theta) = r\exp(i\theta)$, $\theta \in [0, 2\pi]$. By the Fundamental Theorem of Contour Integration, we have

$$0 = \int_C \exp z \, dz = \int_0^{2\pi} e^{r\cos\theta + ir\sin\theta} \cdot ri\exp(i\theta)d\theta$$

$$= \int_0^{2\pi} e^{r\cos\theta} \cdot r \cdot i \cdot \exp(i(r\sin\theta + \theta))d\theta.$$

So we obtain in particular (by equating real parts) that

$$\int_0^{2\pi} e^{r\cos\theta}\cos(r\sin\theta + \theta)d\theta = 0.$$

Solution to Exercise 3.16

Suppose F is holomorphic in $\mathbb{C}\setminus\{0\}$ and that $F' = 1/z$ in $\mathbb{C}\setminus\{0\}$. Consider a circular path C with a positive radius centered at 0 traversed in the anticlockwise direction. Then by the Fundamental Theorem of Contour Integration, since C is closed, we have

$$\int_C F'(z)dz = 0.$$

On the other hand, we know that

$$\int_C F'(z)dz = \int_C \frac{1}{z}dz = 2\pi i,$$

a contradiction.

Solution to Exercise 3.17

(ER1) Let $\gamma : [0, 1] \to D$ be a path which is closed. Define the map $H : [0, 1] \times [0, 1] \to D$ by $H(t, s) = \gamma(t)$, $t, s \in [0, 1]$. Then H is continuous, and

$$H(t, 0) = \gamma(t), \text{ for all } t \in [0, 1],$$
$$H(t, 1) = \gamma(t), \text{ for all } t \in [0, 1],$$
$$H(0, s) = \gamma(0) = \gamma(1) = H(1, s), \text{ for all } s \in [0, 1].$$

Hence γ is D-homotopic to itself. So the relation is reflexive.

(ER2) Let $\gamma_0, \gamma_1 : [0,1] \to D$ be closed paths such that γ_0 is D-homotopic to γ_1. Then there exists a continuous $H : [0,1] \times [0,1] \to D$ such that
$$H(t,0) = \gamma_0(t), \text{ for all } t \in [0,1],$$
$$H(t,1) = \gamma_1(t), \text{ for all } t \in [0,1],$$
$$H(0,s) = H(1,s), \text{ for all } s \in [0,1].$$
Let $\widetilde{H} : [0,1] \times [0,1] \to D$ be defined by $\widetilde{H}(t,s) = H(t,1-s)$ for all $t, s \in [0,1]$. Then $\widetilde{H}$ is continuous and
$$\widetilde{H}(t,0) = H(t,1) = \gamma_1(t), \text{ for all } t \in [0,1],$$
$$\widetilde{H}(t,1) = H(t,0) = \gamma_0(t), \text{ for all } t \in [0,1],$$
$$\widetilde{H}(0,s) = H(0,1-s) = H(1,1-s) = \widetilde{H}(1,s), \text{ for all } s \in [0,1].$$
Thus γ_1 is D-homotopic to γ_0. Hence the relation is symmetric.

(ER3) Let $\gamma_0, \gamma_1, \gamma_2$ be closed paths such that γ_0 is D-homotopic to γ_1 and γ_1 is D-homotopic to γ_2. So there exist two continuous functions $H, K : [0,1] \times [0,1] \to D$ such that
$$H(t,0) = \gamma_0(t), \text{ for all } t \in [0,1], \quad K(t,0) = \gamma_1(t), \text{ for all } t \in [0,1],$$
$$H(t,1) = \gamma_1(t), \text{ for all } t \in [0,1], \quad K(t,1) = \gamma_2(t), \text{ for all } t \in [0,1],$$
$$H(0,s) = H(1,s), \text{ for all } s \in [0,1], \quad K(0,s) = K(1,s), \text{ for all } s \in [0,1].$$
Let $L : [0,1] \times [0,1] \to D$ be defined by
$$L(t,s) = \begin{cases} H(t,2s) & s \in [0, \tfrac{1}{2}], \\ K\left(t, 2(s - \tfrac{1}{2})\right) & s \in (\tfrac{1}{2}, 1] \end{cases}$$
Then
$$L(t,0) = H(t,0) = \gamma_0(t), \text{ for all } t \in [0,1],$$
$$L(t,1) = K(t,1) = \gamma_2(t), \text{ for all } t \in [0,1].$$
Also, for $0 \leq s \leq \tfrac{1}{2}$, $L(0,s) = H(0,2s) = H(1,2s) = L(1,s)$, and for $\tfrac{1}{2} < s \leq 1$,
$$L(0,s) = K\left(0, 2(s - \tfrac{1}{2})\right) = K\left(1, 2(s - \tfrac{1}{2})\right) = L(1,s).$$
Moreover, if $((t_n, s_n))_{n \in \mathbb{N}}$ is a sequence in $[0,1] \times (\tfrac{1}{2}, 1]$ that converges to $(t_0, \tfrac{1}{2})$, then
$$\lim_{n \to \infty} L(t_n, s_n) = \lim_{n \to \infty} K\left(t_n, 2(s_n - \tfrac{1}{2})\right) = K(t_0, 0) = \gamma_1(t_0)$$
$$= H(t_0, 1) = L\left(t_0, \tfrac{1}{2}\right) = L\left(\lim_{n \to \infty} (t_n, s_n)\right).$$
Hence it follows that L is continuous. Consequently γ_0 is D-homotopic to γ_2. So the relation is transitive.

As the relation of D-homotopy is reflexive, symmetric and transitive, it is an equivalence relation.

Solution to Exercise 3.18

From the picture, we see that C is $\mathbb{C} \setminus \{0\}$-homotopic to S.

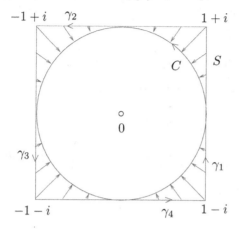

Let

$$\gamma_1(t) := (1-t)(1-i) + t(1+i) = 1 + i(2t-1),$$
$$\gamma_2(t) := (1-t)(1+i) + t(-1+i) = (1-2t) + i,$$
$$\gamma_3(t) := (1-t)(-1+i) + t(-1-i) = -1 + i(1-2t),$$
$$\gamma_4(t) := (1-t)(-1-i) + t(1-i) = 2t - 1 - i,$$

for $t \in [0, 1]$. Then

$$\int_S \frac{1}{z}dz = \int_{\gamma_1} \frac{1}{z}dz + \int_{\gamma_2} \frac{1}{z}dz + \int_{\gamma_3} \frac{1}{z}dz + \int_{\gamma_4} \frac{1}{z}dz.$$

We have

$$
\begin{aligned}
\int_{\gamma_1} \frac{1}{z}dz &= \int_0^1 \frac{2i}{1 + i(2t-1)}dt = \int_0^1 \frac{2i(1 - i(2t-1))}{1 + (2t-1)^2}dt \\
&= 2i\int_0^1 \frac{1}{1 + (2t-1)^2}dt + 2\int_0^1 \frac{2t-1}{1 + (2t-1)^2}dt \\
&\overset{(u=2t-1)}{=} i\int_{-1}^1 \frac{1}{1 + u^2}dt + \int_{-1}^1 \frac{u}{1 + u^2}dt \\
&= i(\tan^{-1}1 - \tan^{-1}(-1)) + 0 = i\left(\frac{\pi}{4} - \left(-\frac{\pi}{4}\right)\right) = i\frac{\pi}{2}.
\end{aligned}
$$

Similarly, using the facts that

$$\int_0^1 \frac{2}{1 + (2t-1)^2}dt = \frac{\pi}{2} \quad \text{and} \quad \int_0^1 \frac{2t-1}{1 + (2t-1)^2}dt = 0,$$

we obtain

$$\int_{\gamma_2} \frac{1}{z} dz = \int_0^1 \frac{-2}{1 - 2t + i} dt = \int_0^1 \frac{-2 \cdot (-(2t - 1) - i)}{1 + (2t - 1)^2} dt$$

$$= 0 + (-1)(-i)\frac{\pi}{2} = i\frac{\pi}{2},$$

$$\int_{\gamma_3} \frac{1}{z} dz = \int_0^1 \frac{-2i}{-1 + i(1 - 2t)} dt = \int_0^1 \frac{-2i \cdot (-1 + i(2t - 1))}{1 + (2t - 1)^2} dt$$

$$= -i \cdot (-1) \cdot \frac{\pi}{2} + 0 = i\frac{\pi}{2}, \text{ and}$$

$$\int_{\gamma_4} \frac{1}{z} dz = \int_0^1 \frac{2}{2t - 1 - i} dt = \int_0^1 \frac{2 \cdot ((2t - 1) + i)}{1 + (2t - 1)^2} dt$$

$$= 0 + i\frac{\pi}{2} = i\frac{\pi}{2}.$$

Thus $\int_S \frac{1}{z} dz = 4 \cdot \left(i\frac{\pi}{2} \right) = 2\pi i$, as expected.

Solution to Exercise 3.19

For a circular path C with center 0 and a positive radius, traversed in the anticlockwise direction, we have

$$\int_C \frac{1}{z} dz = 2\pi i.$$

But the elliptic path E is $\mathbb{C} \setminus \{0\}$-homotopic to C, as shown in Figure 5.17.

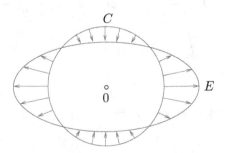

Fig. 5.17 E, C are $\mathbb{C} \setminus \{0\}$-homotopic.

So by the Cauchy Integral Theorem, $\int_E \frac{1}{z}dz = \int_C \frac{1}{z}dz = 2\pi i$, and so

$$2\pi i = \int_E \frac{1}{z}dz = \int_0^{2\pi} \frac{1}{a\cos\theta + ib\sin\theta} \cdot (-a\sin\theta + ib\cos\theta)d\theta$$

$$= \int_0^{2\pi} \frac{(-a\sin\theta + ib\cos\theta)(a\cos\theta - ib\sin\theta)}{a^2(\cos\theta)^2 + b^2(\sin\theta)^2}d\theta$$

$$= \int_0^{2\pi} \frac{(b^2 - a^2)(\cos\theta)(\sin\theta) + iab((\cos\theta)^2 + (\sin\theta)^2)}{a^2(\cos\theta)^2 + b^2(\sin\theta)^2}d\theta$$

$$= \int_0^{2\pi} \frac{(b^2 - a^2)(\cos\theta)(\sin\theta) + iab\cdot 1}{a^2(\cos\theta)^2 + b^2(\sin\theta)^2}d\theta.$$

Equating the imaginary parts, $\int_0^{2\pi} \frac{1}{a^2(\cos\theta)^2 + b^2(\sin\theta)^2}d\theta = \frac{2\pi}{ab}.$

Solution to Exercise 3.20

See Figure 5.18.

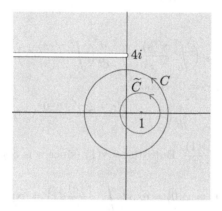

Fig. 5.18

(1) $z \mapsto \mathrm{Log}(z - 4i)$ is holomorphic in $\mathbb{C} \setminus \{r + 4i : r \leq 0\}$. So by the Cauchy Integral Theorem

$$\int_C \mathrm{Log}(z - 4i)dz = 0.$$

(2) If $\widetilde{C}$ denotes the circle with center 1 and any positive radius, say, r, then we know that

$$\int_{\widetilde{C}} \frac{1}{z - 1}dz = 2\pi i.$$

Since $1/(\cdot-1)$ is holomorphic in $\mathbb{C}\setminus\{1\}$, and since the two circular paths C and $\widetilde{C}$ are $\mathbb{C}\setminus\{1\}$-homotopic, it follows from the Cauchy Integral Theorem that

$$\int_C \frac{1}{z-1}dz = \int_{\widetilde{C}} \frac{1}{z-1}dz = 2\pi i.$$

(3) We have

$$i^{z-3} = \exp((z-3)\mathrm{Log}\,i) = \exp\left((z-3)\left(\log 1 + i\frac{\pi}{2}\right)\right)$$

$$= \exp\left((z-3)\left(0 + i\frac{\pi}{2}\right)\right)$$

$$= \exp\left(i\frac{\pi}{2}\cdot(z-3)\right),$$

and so $z \mapsto i^{z-3}$ is entire. By the Cauchy Integral Theorem,

$$\int_C i^{z-3}dz = 0.$$

Solution to Exercise 3.21

(1) We have

$$\varphi'(t) = \exp\left(\int_0^t \frac{\gamma'(s)}{\gamma(s)}ds\right) \cdot \frac{d}{dt}\left(\int_0^t \frac{\gamma'(s)}{\gamma(s)}ds\right) = \varphi(t) \cdot \frac{\gamma'(t)}{\gamma(t)},$$

and so $\varphi'\gamma - \varphi\gamma' = 0$. Hence

$$\frac{d}{dt}\left(\frac{\varphi}{\gamma}\right) = \frac{\varphi'\gamma - \varphi\gamma'}{\gamma^2} = \frac{0}{\gamma^2} = 0,$$

and so $\dfrac{\varphi(0)}{\gamma(0)} = \dfrac{\varphi(1)}{\gamma(1)}$. But $\gamma(0) = \gamma(1)$ (since γ is closed). Hence

$$\varphi(1) = \varphi(0) = \exp\left(\int_0^0 \frac{\gamma'(s)}{\gamma(s)}ds\right) = \exp(0) = 1.$$

Consequently, $w(\gamma) \in \mathbb{Z}$.

(2) We have

$$w(\Gamma_1) = \frac{1}{2\pi i}\int_0^1 \frac{\Gamma_1'(t)}{\Gamma_1(t)}dt$$

$$= \frac{1}{2\pi i}\int_0^1 \frac{2\pi i \exp(2\pi i t)}{\exp(2\pi i t)}dt$$

$$= \frac{1}{2\pi i}\cdot 2\pi i = 1.$$

(3) We have $(\gamma_1 \cdot \gamma_2)'(t) = \gamma_1'(t) \cdot \gamma_2(t) + \gamma_1(t)\gamma_2'(t)$, $t \in [0,1]$, and so

$$
\begin{aligned}
w(\gamma_1 \cdot \gamma_2) &= \frac{1}{2\pi i} \int_0^1 \frac{(\gamma_1 \cdot \gamma_2)'(t)}{(\gamma_1 \cdot \gamma_2)(t)} dt \\
&= \frac{1}{2\pi i} \int_0^1 \frac{\gamma_1'(t) \cdot \gamma_2(t) + \gamma_1(t)\gamma_2'(t)}{\gamma_1(t) \cdot \gamma_2(t)} dt \\
&= \frac{1}{2\pi i} \int_0^1 \frac{\gamma_1'(t) \cdot \cancel{\gamma_2(t)}}{\gamma_1(t) \cdot \cancel{\gamma_2(t)}} dt + \frac{1}{2\pi i} \int_0^1 \frac{\cancel{\gamma_1(t)}\gamma_2'(t)}{\cancel{\gamma_1(t)} \cdot \gamma_2(t)} dt \\
&= w(\gamma_1) + w(\gamma_2).
\end{aligned}
$$

(4) We have $\Gamma_m = \Gamma_1 \cdots \cdots \Gamma_1$ (m times), and so

$$
\begin{aligned}
w(\Gamma_m) &= w(\Gamma_1) + \cdots + w(\Gamma_1) \ (m \text{ times}) \\
&= m \cdot w(\Gamma_1) = m \cdot 1 = m.
\end{aligned}
$$

(5) Consider the map $\varphi : [0,1] \to \mathbb{R}$ given by $\varphi(t) = |\gamma_0(t)|$, $t \in [0,1]$, which measures the distance of $\gamma(t)$ from 0. Being a continuous function, it has a minimum value d_0, and $d_0 > 0$ since γ_0 does not pass through 0. Take $\delta = d_0/2 > 0$. Let γ be a smooth closed path such that

$$
\|\gamma - \gamma_0\|_\infty := \max_{t \in [0,1]} |\gamma(t) - \gamma_0| < \delta.
$$

We will show that γ is $\mathbb{C} \setminus \{0\}$-homotopic to γ_0. Define the function $H : [0,1] \times [0,1] \to \mathbb{C} \setminus \{0\}$ by $H(t,s) = (1-s)\gamma_0(t) + s\gamma(t)$, $t, s \in [0,1]$. Then H is continuous,

$$
\begin{aligned}
&H(t,0) = \gamma_0(t) \text{ for all } t \in [0,1], \\
&H(t,1) = \gamma(t) \text{ for all } t \in [0,1], \text{ and} \\
&H(0,s) = (1-s)\gamma_0(0) + s\gamma(0) \\
&\qquad = (1-s)\gamma_0(1) + s\gamma(1) = H(1,s) \text{ for all } s \in [0,1].
\end{aligned}
$$

Also we note that $H(t,s)$ is never 0, because it is a convex combination of $\gamma_0(t)$ and $\gamma(t)$, and if $(1-s)\gamma_0(t) + s\gamma(t) = 0$ for some t, s, then we arrive at a contradiction. See Figure 5.19.

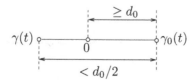

Fig. 5.19 That $H(t,s)$ is never 0.

Indeed,

$$1 \cdot \frac{d_0}{2} > s|\gamma_0(t) - \gamma(t)| = |\gamma_0(t) - ((1-s)\gamma_0(t) + s\gamma(t))| = |\gamma_0(t) - 0|$$
$$= |\gamma_0(t)| \geq d_0.$$

Thus by the Cauchy Integral Theorem,

$$w(\gamma) = \frac{1}{2\pi i} \int_\gamma \frac{1}{z} dz = \frac{1}{2\pi i} \int_\gamma \frac{1}{z} dz = w(\gamma_0).$$

Solution to Exercise 3.22

(1)

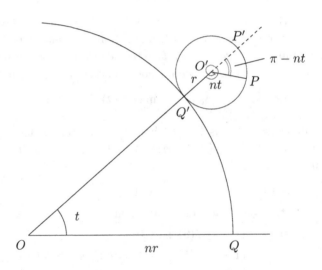

The length of the arc on the big circle subtended by $\angle Q'OQ$ is $t \cdot nr$. As the small coin rolls without slipping, the angle made by $O'P$ with OO' is $(t \cdot nr)/r = n \cdot t$. From the picture, we see that $O' \equiv (nr+r)\exp(it)$, and

$$P \equiv (nr+r)\exp(it) + \underbrace{\exp(-i(\pi - nt))}_{\substack{\text{produces a} \\ \text{clockwise rotation}}} \cdot \underbrace{r\exp(it)}_{O'P'}$$

$$= (n+1)r\exp(it) + (-1) \cdot r \cdot \exp((n+1)it).$$

(2) The area enclosed by the epicycloid γ is

$$\frac{1}{2i}\int_\gamma \overline{z}\,dz = \frac{1}{2i}\int_0^{2\pi}\overline{r\Big((n+1)\exp(it) - \exp((n+1)it)\Big)} \cdot$$

$$r\Big((n+1)i\exp(it) - (n+1)i\exp((n+1)it)\Big)dt$$

$$= \frac{1}{2i}\int_0^{2\pi} r^2\Big((n+1)\exp(-it) - \exp(-(n+1)it)\Big) \cdot$$

$$(n+1)i\Big(\exp(it) - \exp((n+1)it)\Big)dt$$

$$= \frac{(n+1)r^2}{2}\int_0^{2\pi}\big((n+1) - (n+1)\exp(int) - \exp(-int) + 1\big)dt$$

$$= \frac{(n+1)r^2}{2}\big((n+1)\cdot 2\pi + 0 + 0 + 2\pi\big)$$

$$= (n+1)r^2\pi(n+2) = \pi r^2(n+1)(n+2).$$

Solution to Exercise 3.23

Take for example f given by $f(z) = 1/z$ for $z \in D := \mathbb{C} \setminus \{0\}$. Then we have seen that f does not have a primitive in D. (See Example 3.7 and Exercise 3.16.)

Solution to Exercise 3.24

Let $\gamma(t) = \exp(it)$, $t \in [0,1]$ and so

$$\int_\gamma \frac{i}{(z-a)(az-1)}\,dz = \int_0^{2\pi} \frac{i}{(\exp(it) - a)(a\exp(it) - 1)}\,i\exp(it)\,dt$$

$$= \int_0^{2\pi} \frac{-\exp(it)}{(\exp(it) - a)(a - \exp(-it))\exp(it)}\,dt$$

$$= \int_0^{2\pi} \frac{1}{(\exp(it) - a)(\exp(-it) - a)}\,dt$$

$$= \int_0^{2\pi} \frac{1}{|\exp(it) - a|^2}\,dt$$

$$= \int_0^{2\pi} \frac{1}{((\cos t) - a)^2 + (\sin t)^2}\,dt$$

$$= \int_0^{2\pi} \frac{1}{1 - 2a\cos t + a^2}\,dt.$$

Since the mapping $z \mapsto i/(az - 1)$ is holomorphic in a disc containing the unit circle γ (because $0 < a < 1$), by the Cauchy Integral Formula,

$$\frac{1}{2\pi i} \int_{\gamma} \frac{\frac{i}{az-1}}{z - a}\, dz = \frac{i}{az - 1}\bigg|_{z=a} = \frac{i}{a^2 - 1}.$$

So $\displaystyle\int_0^{2\pi} \frac{1}{1 - 2a\cos t + a^2}\, dt = \int_{\gamma} \frac{\frac{i}{az-1}}{z - a}\, dz = 2\pi i \cdot \frac{i}{a^2 - 1} = \frac{2\pi}{1 - a^2}.$

Solution to Exercise 3.25

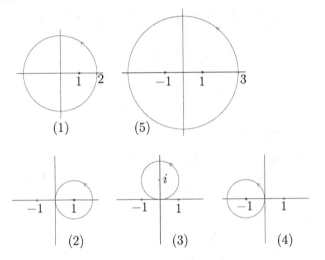

(1)

(5)

(2)

(3)

(4)

(1) $\displaystyle\int_{\gamma} \frac{\exp z}{z - 1}\, dz = 2\pi i \exp z\bigg|_{z=1} = 2\pi i \exp 1 = 2\pi i e.$

(2) $\displaystyle\int_{\gamma} \frac{z^2 + 1}{z^2 - 1}\, dz = \int_{\gamma} \frac{\frac{z^2+1}{z+1}}{z - 1}\, dz = 2\pi i \frac{z^2 + 1}{z + 1}\bigg|_{z=1} = 2\pi i \frac{1^2 + 1}{1 + 1} = 2\pi i.$

(3) $\displaystyle\int_{\gamma} \frac{z^2 + 1}{z^2 - 1}\, dz = 0.$

(4) $\displaystyle\int_{\gamma} \frac{z^2 + 1}{z^2 - 1}\, dz = \int_{\gamma} \frac{\frac{z^2+1}{z-1}}{z - (-1)}\, dz = 2\pi i \frac{z^2 + 1}{z - 1}\bigg|_{z=-1} = 2\pi i \frac{(-1)^2 + 1}{-1 - 1} = -2\pi i.$

(5) $\displaystyle\int_{\gamma} \frac{z^2 + 1}{z^2 - 1}\, dz = \int_{\gamma} \frac{z^2 + 1}{2}\left(\frac{1}{z - 1} - \frac{1}{z + 1}\right) dz$

$\displaystyle\qquad = \int_{\gamma} \frac{\frac{z^2+1}{2}}{z - 1}\, dz - \int_{\gamma} \frac{\frac{z^2+1}{2}}{z - (-1)}\, dz = 2\pi i \frac{z^2 + 1}{2}\bigg|_{z=1} - 2\pi i \frac{z^2 + 1}{2}\bigg|_{z=-1}$

$\displaystyle\qquad = 2\pi i(1) - 2\pi i(1) = 0.$

Solution to Exercise 3.26

Suppose F is a primitive. Consider the closed path γ given by $|z - 0| = \frac{1}{2}$ traversed anticlockwise. By the Fundamental Theorem of Contour Integration,

$$\int_\gamma \frac{1}{z(z^2 - 1)} dz = \int_\gamma F'(z) dz = 0,$$

since γ is closed. On the other hand, by the Cauchy Integral Formula,

$$\int_\gamma \frac{1}{z(z^2 - 1)} dz = \int_\gamma \frac{\frac{1}{z^2-1}}{z - 0} dz = 2\pi i \frac{1}{z^2 - 1} \Big|_{z=0} = 2\pi i \frac{1}{0^2 - 1} = -2\pi i.$$

So we arrive at a contradiction. Hence

$$\frac{1}{z(z^2 - 1)}$$

does not possess a primitive in $\{z \in \mathbb{C} : 0 < |z| < 1\}$.

Solution to Exercise 3.27

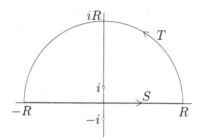

Fig. 5.20   The path $\sigma = S + T$.

(1) By the Cauchy Integral Formula, we have

$$\int_\sigma F(z) dz = \int_\sigma \frac{\exp(iz)}{z^2 + 1} dz = \int_\sigma \frac{\frac{\exp(iz)}{z+i}}{z - i} dz = 2\pi i \frac{\exp(iz)}{z + i} \Big|_{z=i}$$

$$= 2\pi i \frac{\exp(i \cdot i)}{i + i} = 2\pi i \frac{e^{-1}}{2i} = \frac{\pi}{e}.$$

(2) Let $z = z + iy$, with x, y real and $y \geq 0$. Then

$$|\exp(iz)| = |\exp(i(x + iy))| = |\exp(-y + ix)| = e^{-y} \leq 1$$

(since $y \geq 0$). Thus

$$|F(z)| = \frac{|\exp(iz)|}{|z^2 + 1|} \leq \frac{1}{|z^2 + 1|}.$$

But $|z^2| - |-1| \leq |z^2 - (-1)| = |z^2 + 1|$, and so if $|z| \geq \sqrt{2}$,

$$|F(z)| \leq \frac{1}{|z^2 + 1|} \leq \frac{1}{|z|^2 - 1} \leq \frac{2}{|z|^2}$$

since $|z|^2 \leq 2|z|^2 - 2$ for $|z^2| \geq 2$, that is, $|z| \geq \sqrt{2}$.

(3) We have

$$\left| \int_T F(z)dz \right| \leq 2\pi R \cdot \max_{z \in T} |F(z)| \leq 2\pi R \cdot \frac{2}{R^2} \quad \text{(for } R \geq \sqrt{2})$$

$$= \frac{4\pi}{R} \xrightarrow{R \to \infty} 0.$$

So $\lim\limits_{R \to \infty} \int_T F(z)dz = 0$. Since

$$\int_S F(z)dz = \int_\sigma F(z)dz - \int_T F(z)dz = \frac{\pi}{e} - \int_T F(z)dz,$$

it now follows that $\lim\limits_{R \to \infty} \int_S F(z)dz = \frac{\pi}{e} - \lim\limits_{R \to \infty} \int_T F(z)dz = \frac{\pi}{e} - 0 = \frac{\pi}{e}$.

(4) Let $S(x) = x$, $x \in [-R, R]$. Then

$$\int_S F(z)dz = \int_{-R}^R \frac{\exp(ix)}{x^2 + 1} \cdot 1 \, dx = \int_{-R}^R \frac{\cos x}{x^2 + 1} dx + i \int_{-R}^R \frac{\sin x}{x^2 + 1} dx$$

$$= \int_{-R}^R \frac{\cos x}{x^2 + 1} dx + 0$$

where we have used the fact that $\dfrac{\sin x}{x^2 + 1}$ is an odd function to get the last equality. Hence

$$\lim_{R \to \infty} \int_{-R}^R \frac{\cos x}{x^2 + 1} dx = \lim_{R \to \infty} \int_S F(z)dz = \frac{\pi}{e}.$$

Solution to Exercise 3.28

Consider the entire function $\exp z$ and let $C : [0, 2\pi] \to \mathbb{C}$ be the circular path with center 0 and radius 1 given by $C(\theta) = \exp(i\theta)$, $\theta \in [0, 2\pi]$. By the Cauchy Integral Formula,

$$\frac{1}{2\pi i} \int_C \frac{\exp z}{z - 0} dz = \exp z \Big|_{z=0} = \exp 0 = 1,$$

But

$$\int_C \frac{\exp z}{z - 0}\, dz = \int_0^{2\pi} \frac{\exp(\exp(i\theta))}{\exp(i\theta)} \cdot i\exp(i\theta)\, d\theta = i \int_0^{2\pi} \exp(\exp(i\theta))\, d\theta$$

$$= i \int_0^{2\pi} \exp(\cos\theta + i\sin\theta)\, d\theta$$

$$= i \int_0^{2\pi} e^{\cos\theta}(\cos(\sin\theta) + i\sin(\sin\theta))\, d\theta$$

$$= -\int_0^{2\pi} e^{\cos\theta}\sin(\sin\theta)\, d\theta + i \int_0^{2\pi} e^{\cos\theta}\cos(\sin\theta)\, d\theta$$

Hence $\displaystyle\int_0^{2\pi} e^{\cos\theta}\cos(\sin\theta)\, d\theta = 2\pi$.

Solution to Exercise 3.29

If f is holomorphic, then $f^{(n)}$ is holomorphic too, and so is its derivative $f^{(n+1)}$. But $f^{(n+1)}$, being complex differentiable, is in particular continuous. So $f^{(n)}$ has a continuous complex derivative.

Solution to Exercise 3.30

Since for all $z \in \mathbb{C}$, $|f(z)| \geq \delta > 0$, in particular $f(z) \neq 0$ for all $z \in \mathbb{C}$, and so $1/f$ is entire. But

$$\text{for all } z \in \mathbb{C} \quad \left|\frac{1}{f(z)}\right| \leq \frac{1}{\delta},$$

and so by Liouville's Theorem, $1/f$ is constant. Thus f is constant as well.

Solution to Exercise 3.31

Let g be defined by $g(z) := f(z) - w_0$ for $z \in \mathbb{C}$. Then g is entire and $|g(z)| \geq r$ for all $z \in \mathbb{C}$, and so g is bounded away from 0. Hence by Exercise 3.30, g is constant, and so $f = g + w_0$ must be constant too.

Solution to Exercise 3.32

Consider the compact set $K := \{(x, y) : 0 \leq x \leq T_1,\ 0 \leq y \leq T_2\}$. The continuous function $(x, y) \mapsto |f(x + iy)|$ assumes a maximum value M on

K. But for

$$x, y \in \mathbb{R} = \bigcup_{n \in \mathbb{Z}} [nT_1, (n+1)T_1) = \bigcup_{m \in \mathbb{Z}} [mT_1, (m+1)T_1),$$

there are integers n, m such that $x + iy = x_0 + nT_1 + i(y_0 + mT_2)$ for some $x_0 \in [0, T_1)$ and $y_0 \in [0, T_2)$. Owing to the periodicity of f,

$$f(x+iy) = f(x_0 + nT_1 + i(y_0 + mT_2)) = f(x_0 + iy_0) \in f(K),$$

and so for all $x, y \in \mathbb{R}$, $|f(x_0 + iy_0)| \leq M$. Hence f is bounded on $\mathbb{C}$, and by Liouville's Theorem, must be constant.

Solution to Exercise 3.33

(1) Let g be defined by $g(z) = \exp(-z) \cdot f(z)$ for $z \in \mathbb{C}$. Then g is entire. Moreover, since $|f(z)| \leq |\exp z|$, it follows by rearranging that

$$|g(z)| = |\exp(-z) \cdot f(z)| \leq 1$$

for all $z \in \mathbb{C}$. So by Liouville's Theorem, g is constant with value say, c. Since $|g(z)| \leq 1$, we obtain $|c| \leq 1$. Consequently,

$$g(z) = \exp(-z) \cdot f(z) = c,$$

and so $f(z) = c \cdot \exp z$ for all $z \in \mathbb{C}$, where c is a constant such that $|c| \leq 1$.

(2) We know that if p is a polynomial of degree $d \geq 1$, then there exist $M, R > 0$ such that

$$|p(z)| \geq M|z|^d$$

for $|z| > R$. Thus with $z = x < -R < 0$, we have $|z| > R$ and so $M|x|^d \leq |p(z)| \leq |e^x| = e^x \leq 1$ (because $x < 0$). Hence $|x|^d \leq 1/M$ for all $x < -R$, a contradiction. So p is a constant, say equal to c_0. But then $|p(z)| \leq |\exp z|$ again gives with $z = x < 0$ that $|c_0| \leq |e^x| = e^x$ for all $x < 0$, and so $|c_0| = 0$. Consequently, $p = c_0 = 0$.

Solution to Exercise 3.34

(1) We have for $z \in C$ that

$$|z - a_1| \geq |z| - |a_1| = R - |a_1|,$$
$$|z - a_2| \geq |z| - |a_2| = R - |a_2|.$$

Thus

$$\left| \int_C \frac{f(z)}{(z-a_1)(z-a_2)} dz \right| \leq \max_{z \in C} \frac{|f(z)|}{|z-a_1||z-a_2|} \cdot (\text{length of } C)$$

$$\leq \frac{M}{(R-|a_1|)(R-|a_2|)} \cdot 2\pi R,$$

where $M := \max_{z \in C} |f(z)|$.

(2) Since $a_1 \neq a_2$, we have

$$\frac{1}{z-a_1} - \frac{1}{z-a_2} = \frac{z - a_2 - (z-a_1)}{(z-a_1)(z-a_2)} = \frac{a_1 - a_2}{(z-a_1)(z-a_2)},$$

and so

$$\frac{1}{(z-a_1)(z-a_2)} = \frac{1}{a_1 - a_2} \left(\frac{1}{z-a_1} - \frac{1}{z-a_2} \right).$$

Hence $\alpha := -\beta := \dfrac{1}{a_1 - a_2}$.

(3) We have

$$\int_C \frac{f(z)}{(z-a_1)(z-a_2)} dz = \int_C \frac{1}{a_1 - a_2} \left(\frac{f(z)}{z-a_1} - \frac{f(z)}{z-a_2} \right) dz$$

$$= \frac{1}{a_1 - a_2} \left(\int_C \frac{f(z)}{z-a_1} dz - \int_C \frac{f(z)}{z-a_2} dz \right).$$

If we consider a small disc Δ_1 with center a_1 and radius $r_1 > 0$ with boundary C_1, then we see that C and C_1 are $\mathbb{C} \setminus \{a_1\}$-homotopic, and the function

$$g(z) := \frac{f(z)}{z-a_1}, \quad z \in \mathbb{C} \setminus \{a_1\}$$

is holomorphic. Thus by the Cauchy Integral Theorem,

$$\int_C \frac{f(z)}{z-a_1} dz = \int_{C_1} \frac{f(z)}{z-a_1} dz.$$

But by the Cauchy Integral Formula, $\dfrac{1}{2\pi i} \displaystyle\int_{C_1} \dfrac{f(z)}{z-a_1} dz = f(a_1)$. Thus

$$\int_C \frac{f(z)}{z-a_1} dz = 2\pi i f(a_1).$$

Similarly,

$$\int_C \frac{f(z)}{z-a_2} dz = 2\pi i f(a_2).$$

Consequently, $\displaystyle\int_C \frac{f(z)}{(z-a_1)(z-a_2)} dz = \frac{2\pi i (f(a_1) - f(a_2))}{a_1 - a_2}.$

(4) Suppose that f is a bounded entire function with a bound M on $|f|$, that is, $|f(z)| \leq M$ for all $z \in \mathbb{C}$. Suppose that a_1, a_2 are any two distinct points in $\mathbb{C}$. Let C be a circular path with radius $R > 0$ and center 0, traversed once in the counterclockwise direction that contains a_1, a_2 in its interior. Then using the results in parts (i) and (iii),

$$
\begin{aligned}
|f(a_1) - f(a_2)| &= \frac{|a_1 - a_2|}{2\pi} \cdot \left| \frac{2\pi i (f(a_1) - f(a_2))}{a_1 - a_2} \right| \\
&= \frac{|a_1 - a_2|}{2\pi} \cdot \left| \int_C \frac{f(z)}{(z - a_1)(z - a_2)} dz \right| \\
&\leq \frac{|a_1 - a_2|}{2\pi} \cdot \frac{2\pi R M}{(R - |a_1|)(R - |a_2|)}.
\end{aligned}
$$

As R can be made as large as we please and since

$$
\frac{2\pi R M}{(R - |a_1|)(R - |a_2|)} \to 0
$$

as $R \to \infty$, it follows that $|f(a_1) - f(a_2)| = 0$. This implies that $f(a_1) = f(a_2)$. Hence f is constant.

Solutions to the exercises from Chapter 4

Solution to Exercise 4.1

Since $\displaystyle\sum_{n=1}^{\infty} a_n$ converges, so do the real series $\displaystyle\sum_{n=1}^{\infty} \mathrm{Re}(a_n)$ and $\displaystyle\sum_{n=1}^{\infty} \mathrm{Im}(a_n)$.

Hence $\displaystyle\lim_{n\to\infty} \mathrm{Re}(a_n) = 0$ and $\displaystyle\lim_{n\to\infty} \mathrm{Im}(a_n) = 0$. Thus $\displaystyle\lim_{n\to\infty} a_n = 0$ too.

Solution to Exercise 4.2

As $\displaystyle\sum_{n=1}^{\infty} |a_n|$ converges, and for all $n \in \mathbb{N}$, $\mathrm{Re}(a_n) \leq |a_n|$, $\mathrm{Im}(a_n) \leq |a_n|$,

$$\sum_{n=1}^{\infty} \mathrm{Re}(a_n) \text{ and } \sum_{n=1}^{\infty} \mathrm{Im}(a_n)$$

converge by the Comparison Test. Hence $\displaystyle\sum_{n=1}^{\infty} a_n$ converges.

Solution to Exercise 4.3

With $s_n := 1+z+\cdots+z^{n-1}+z^n$, we have that $zs_n = z+z^2+\cdots+z^n+z^{n+1}$, and so $(1 - z)s_n = 1 - z^{n+1}$. Since $|z| < 1$, $z \neq 1$ and so $1 - z \neq 0$. So

$$s_n = 1 + z + \cdots + z^{n-1} + z^n = \frac{1 - z^{n+1}}{1 - z}. \tag{5.21}$$

Thus

$$\lim_{n\to\infty} s_n = \lim_{n\to\infty} \frac{1 - z^{n+1}}{1 - z} = \frac{1 - 0}{1 - z} = \frac{1}{1 - z},$$

and so $\displaystyle\sum_{n=0}^{\infty} z^n$ converges, with $\displaystyle\sum_{n=0}^{\infty} z^n = \lim_{n\to\infty} s_n = \frac{1}{1 - z}$.
(Note that we have used the fact that since $|z| < 1$, one has

$$\lim_{n\to\infty} z^{n+1} = 0,$$

and this can be justified as follows: $|z^{n+1} - 0| = |z|^{n+1} \overset{n\to\infty}{\longrightarrow} 0$, since $r := |z| < 1$.)

Solution to Exercise 4.4

For $n \in \mathbb{N}$, let $s_n := 1 + 2z + 3z^2 + \cdots + (n-1)z^{n-2} + nz^{n-1}$. Then $zs_n = z + 2z^2 + \cdots + (n-1)z^{n-1} + nz^n$. Hence

$$(1-z)s_n = 1 + z + z^2 + \cdots + z^{n-1} - nz^n = \frac{1-z^n}{1-z} - nz^n.$$

Thus

$$s_n = \frac{1-z^n}{(1-z)^2} - \frac{nz^n}{1-z}.$$

(We could have also obtained this expression from (5.21) by differentiating with respect to z.) If we set $r := |z|$, then $0 \le r < 1$ and so

$$r = \frac{1}{1+h}$$

where $h := \dfrac{1}{r} - 1 > 0$. We have

$$(1+h)^n = 1 + \binom{n}{1}h + \binom{n}{2}h^2 + \cdots + \binom{n}{n}h^n \ge \binom{n}{2}h^2 = \frac{n \cdot (n-1)}{2} \cdot h^2.$$

Hence

$$0 \le nr^n = \frac{n}{(1+h)^n} \le n \cdot \frac{2}{n \cdot (n-1) \cdot h^2} = \frac{2}{(n-1) \cdot h^2},$$

and so by the Sandwich Theorem, $\lim\limits_{n\to\infty} nr^n = 0$. Consequently,

$$\lim_{n\to\infty} s_n = \lim_{n\to\infty}\left(\frac{1-z^n}{(1-z)^2} - \frac{nz^n}{1-z} \right) = \frac{1-0}{(1-z)^2} - \frac{0}{1-z} = \frac{1}{(1-z)^2}.$$

Solution to Exercise 4.5

We have

$$\left| \frac{1}{n^s} \right| = \left| \frac{1}{\exp(s \cdot \mathrm{Log}(n))} \right| = \left| \frac{1}{\exp(s \cdot \log n)} \right|$$

$$= \frac{1}{e^{\mathrm{Re}(s \cdot \log n)}} = \frac{1}{e^{(\log n)\cdot(\mathrm{Re}(s))}} = \frac{1}{(e^{\log n})^{\mathrm{Re}(s)}} = \frac{1}{n^{\mathrm{Re}(s)}}.$$

Recall that $\sum\limits_{n=1}^{\infty} \dfrac{1}{n^p}$ converges if $p > 1$. Hence if $\mathrm{Re}(s) > 1$, then

$$\sum_{n=1}^{\infty} \frac{1}{n^{\mathrm{Re}(s)}}$$

converges. Thus

$$\sum_{n=1}^{\infty} \frac{1}{n^s}$$

converges absolutely for $\mathrm{Re}(s) > 1$, and in particular, it converges for $\mathrm{Re}(s) > 1$.

Solution to Exercise 4.6

Let $L \neq 0$. We have that for all z such that $|z| < 1/L$ that there exists a $q < 1$ and an N large enough such that $\sqrt[n]{|c_n z^n|} = \sqrt[n]{|c_n|} \, |z| \leq q < 1$ for all $n > N$. This is because $\sqrt[n]{|c_n|} \, |z| \xrightarrow{n \to \infty} L|z| < 1$. (For example take $q = (L|z| + 1)/2 < 1$.) So by the Root Test, the power series converges absolutely for such z.

If $L = 0$, then for any $z \in \mathbb{C}$, we can guarantee that there exists a $q < 1$ such that $\sqrt[n]{|c_n z^n|} = \sqrt[n]{|c_n|} \, |z| \leq q < 1$ for all $n > N$. This is because $\sqrt[n]{|c_n|} \, |z| \xrightarrow{n \to \infty} 0|z| = 0 < 1$. (So we may arrange for example that $q = 1/2 < 1$.) So again by the Root Test, the power series converges absolutely for such z.

On the other hand, if $L \neq 0$ and $|z| > 1/L$, then there exists an N large enough such that $\sqrt[n]{|c_n z^n|} = \sqrt[n]{|c_n|} \, |z| > 1$ for all $n > N$. This is because $\sqrt[n]{|c_n|} \, |z| \xrightarrow{n \to \infty} L|z| > 1$. So again by the Root Test, the power series diverges.

Solution to Exercise 4.7

When $z = 0$, the series clearly converges with sum 0. Suppose $z \neq 0$. Then $|z| \neq 0$. Choose $N \in \mathbb{N}$ such that $N > 1/|z|$. Then for $n > N$, $|nz| > N|z| > 1$, and so $|n^n z^n - 0| = |nz|^n > 1^n = 1$, showing that

$$\neg \left(\lim_{n \to \infty} n^n z^n = 0 \right).$$

Thus if $z \neq 0$, then $\sum_{n=1}^{\infty} n^n z^n$ diverges.

Solution to Exercise 4.8

We have $\lim_{n \to \infty} \sqrt[n]{\dfrac{1}{n^n}} = \lim_{n \to \infty} \dfrac{1}{n} = 0$. So the radius of convergence of

$$\sum_{n=1}^{\infty} \frac{z^n}{n^n}$$

is infinite, and the power series converges for all $z \in \mathbb{C}$.

Solution to Exercise 4.9

(1) We have

$$\lim_{n\to\infty} \left| \frac{\frac{(-1)^{n+1}}{n+1}}{\frac{(-1)^n}{n}} \right| = \lim_{n\to\infty} \frac{n}{n+1} = 1,$$

and so the radius of convergence of $\displaystyle\sum_{n=1}^{\infty} \frac{(-1)^n}{n} z^n$ is 1.

(2) We have

$$\lim_{n\to\infty} \left| \frac{(n+1)^{2012}}{n^{2012}} \right| = \lim_{n\to\infty} \left(1 + \frac{1}{n}\right)^{2012} = 1,$$

and so the radius of convergence of $\displaystyle\sum_{n=0}^{\infty} n^{2012} z^n$ is 1.

(3) We have

$$\lim_{n\to\infty} \left| \frac{\frac{1}{(n+1)!}}{\frac{1}{n!}} \right| = \lim_{n\to\infty} \frac{1}{n+1} = 0,$$

and so the radius of convergence of $\displaystyle\sum_{n=0}^{\infty} \frac{1}{n!} z^n$ is infinite.

Solution to Exercise 4.10

For $|z| < 1$, we know that

$$f(z) := 1 + 2z + 3z^2 + 4z^3 + \cdots = \frac{1}{(1-z)^2},$$

and so

$$zf(z) = g(z) := z + 2z^2 + 3z^3 + 4z^4 + \cdots = \frac{z}{(1-z)^2}$$

for $|z| < 1$. So $g(z) := z + 2z^2 + 3z^3 + 4z^4 + \cdots$ converges for $|z| < 1$, g is holomorphic in the disc $|z| < 1$, with $g'(z) = 1 + 2^2 z + 3^2 z^2 + 4^2 z^3 + \cdots$ for $|z| < 1$. On the other hand,

$$g(z) = zf(z) = \frac{z}{(1-z)^2} \text{ for } |z| < 1,$$

and so

$$g'(z) = \frac{d}{dz}\left(\frac{z}{(1-z)^2}\right) = 1 \cdot \frac{1}{(1-z)^2} + z \cdot \frac{2}{(1-z)^3} = \frac{1-z+2z}{(1-z)^3} = \frac{1+z}{(1-z)^3}.$$

Solution to Exercise 4.11

(1) False. For example, $\left\{ z \in \mathbb{C} : \sum_{n=1}^{\infty} \dfrac{z^n}{n^2} \text{ converges} \right\} = \{ z \in \mathbb{C} : |z| \le 1 \}$.

(2) True.

(3) False. For example $\sum_{n=1}^{\infty} \dfrac{(-1)^n}{n} z^n$ converges for $z = 1$, but diverges for $z = -1$.

(4) False. See the example in (3).

(5) True. Same example as in (3).

(6) True. For example, consider $\sum_{n=1}^{\infty} \dfrac{z^n}{n^2}$.

(7) True. The radius of convergence is ≤ 1 and $|1 + i| = \sqrt{2} > 1$.

Solution to Exercise 4.12

Since
$$\frac{d^{2n}}{dz^{2n}} \sin z = (-1)^n \sin z \text{ and } \frac{d^{2n+1}}{dz^{2n+1}} \sin z = (-1)^n \cos z,$$
and $\sin 0 = 0$ and $\cos 0 = 1$, we have
$$\sin z = \sum_{n=0}^{\infty} \frac{1}{n!} \left(\frac{d^n}{dz^n} \sin z \right) \Big|_{z=0} z^n = z - \frac{z^3}{3!} + \frac{z^5}{5!} - + \cdots.$$
Similarly, $\cos z = 1 - \dfrac{z^2}{2!} + \dfrac{z^4}{4!} - + \cdots$. Alternately,
$$\cos z = \frac{\exp(iz) + \exp(-iz)}{2} = \frac{1}{2} \left(\sum_{n=0}^{\infty} \frac{1}{n!} i^n z^n + \sum_{n=0}^{\infty} \frac{1}{n!} (-1)^n i^n z^n \right).$$
Since $i^{2n} = (-1)^n$, we have
$$\cos z = \frac{1}{2} \Big(1 + iz - \frac{z^2}{2!} - \frac{iz^3}{3!} + \frac{z^4}{4!} + \frac{iz^5}{5!} - \frac{z^6}{6!} + \cdots$$
$$1 - iz - \frac{z^2}{2!} + \frac{iz^3}{3!} + \frac{z^4}{4!} - \frac{iz^5}{5!} - \frac{z^6}{6!} + \cdots \Big)$$
$$= 1 - \frac{1}{2!} z^2 + \frac{1}{4!} z^4 - \frac{1}{6!} z^6 + - \cdots.$$

Solution to Exercise 4.13

Let $p(z) = z^6 - z^4 + z^2 - 1$, $z \in \mathbb{C}$. Then

$$p'(z) = 6z^5 - 4z^3 + 2z,$$
$$p''(z) = 30z^4 - 12z^2 + 2,$$
$$p'''(z) = 120z^3 - 24z,$$
$$p^{(4)}(z) = 360z^2 - 24,$$
$$p^{(5)}(z) = 720z,$$
$$p^{(6)}(z) = 720,$$
$$p^{(7)}(z) = p^{(8)}(z) = \cdots = 0.$$

Hence

$$p(1) = 1 - 1 + 1 - 1 = 0,$$
$$\frac{p'(1)}{1!} = 6 - 4 + 2 = 4,$$
$$\frac{p''(1)}{2!} = \frac{30 - 12 + 2}{2} = 10,$$
$$\frac{p'''(1)}{3!} = \frac{120 - 24}{6} = 16,$$
$$\frac{p^{(4)}(1)}{4!} = \frac{360 - 24}{24} = 14,$$
$$\frac{p^{(5)}(1)}{5!} = \frac{720}{120} = 6,$$
$$\frac{p^{(6)}(1)}{6!} = \frac{720}{720} = 1.$$

Thus for all $z \in \mathbb{C}$,

$$z^6 - z^4 + z^2 - 1$$
$$= p(1) + \frac{p'(1)}{1!}(z-1) + \cdots + \frac{p^{(6)}(1)}{6!}(z-1)^6 + 0$$
$$= 4(z-1) + 10(z-1)^2 + 16(z-1)^3 + 14(z-1)^4 + 6(z-1)^5 + (z-1)^6.$$

Solution to Exercise 4.14

(1) The function $z \mapsto \exp(z^2)$ possesses a primitive, say g, in the simply connected domain $\mathbb{C}$. Thus

$$f(z) = \int_{\gamma_{0z}} \exp(\zeta^2)d\zeta = \int_{\gamma_{0z}} g'(\zeta)d\zeta = g(z) - g(0).$$

So $f'(z) = g'(z) = \exp(z^2) = \sum_{n=0}^{\infty} \frac{1}{n!} z^{2n}$. Consequently,

$$\frac{1}{(2n)!} \frac{d^{2n}}{dz^{2n}} f'(z) \bigg|_{z=0} = \frac{1}{n!} \text{ and } \frac{1}{(2n+1)!} \frac{d^{2n+1}}{dz^{2n+1}} f'(z) \bigg|_{z=0} = 0.$$

Hence $f^{(2n+1)}(0) = \frac{(2n)!}{n!}$ and $f^{(2n+2)}(0) = 0$. Also, $f(0) = 0$. Thus

$$f(z) = \sum_{n=0}^{\infty} \frac{f^{(n)}(0)}{n!} z^n = \sum_{n=0}^{\infty} \frac{f^{(2n+1)}(0)}{(2n+1)!} z^{2n+1} = \sum_{n=0}^{\infty} \frac{1}{(2n+1)(n!)} z^{2n+1}.$$

(2) For $|z| < 1$,

$$\frac{1}{z+1} = 1 - z + z^2 - z^3 + z^4 - + \cdots,$$

and since power series are holomorphic in the region of convergence with complex derivative given by termwise differentiation, we obtain for $|z| < 1$ that

$$-\frac{1}{(z+1)^2} = \frac{d}{dz} \frac{1}{z+1} = -1 + 2z - 3z^2 + 4z^3 - + \cdots.$$

Multiplying both sides by $-z^2$ gives

$$\frac{z^2}{(z+1)^2} = z^2 - 2z^3 - 3z^4 + - \cdots = \sum_{n=2}^{\infty} (-1)^n \cdot (n-1) \cdot z^n$$

for $|z| < 1$. So we have $c_0 = c_1 = 0$ and $c_n = (-1)^n \cdot (n-1)$ for $n \geq 2$.

Solution to Exercise 4.15

Let $z \in \mathbb{C}$. Let $R > |z|$. Then

$$|f^{(n+1)}(z)| \leq \frac{(n+1)!}{R^{n+1}} \cdot \max_{|z| \leq R} |f(z)|$$

$$\leq \frac{(n+1)!}{R^{n+1}} \cdot \max_{|z| \leq R} M \cdot |z|^n = \frac{(n+1)!}{R^{n+1}} \cdot M \cdot R^n = \frac{(n+1)!M}{R}.$$

But the choice of $|z| > R$ was arbitrary, and so $f^{(n+1)}(z) = 0$. Since $z \in \mathbb{C}$ was arbitrary, we obtain that $f^{(n+1)} \equiv 0$ in $\mathbb{C}$. By Taylor's Theorem, for all $z \in \mathbb{C}$,

$$f(z) = \sum_{k=0}^{\infty} \frac{f^{(k)}(0)}{k!} (z-0)^k = \sum_{k=0}^{n} \frac{f^{(k)}(0)}{k!} z^k,$$

since $f^{(n+1)}(0) = f^{(n+2)}(0) = f^{(n+3)}(0) = \cdots = 0$. So we see that f is a polynomial of degree at most n.

If $n = 0$, then f is a bounded entire function, and our conclusion obtained above says that f is constant. So the special case when $n = 0$ is Liouville's Theorem.

Solution to Exercise 4.16

By the Cauchy Integral Formula,

$$\frac{2013!}{2\pi i} \int_C \frac{\sin z}{z^{2013}} dz = \frac{d^{2012}}{dz^{2012}} \sin z \Big|_{z=0}$$

$$= (-1)^{2012/2} \sin z \Big|_{z=0}$$

$$= 0.$$

Hence $\displaystyle \int_C \frac{\sin z}{z^{2013}} dz = 0$.

Solution to Exercise 4.17

By the continuity of g at z_0, there exists a $\delta > 0$, which can be chosen smaller than R, such that $g(z) \neq 0$ for $|z - z_0| < \delta$. We have $f(z_0) = 0$, but $f(z) = (z - z_0)^m g(z)$ $(|z - z_0| < R)$ shows that $f(z) \neq 0$ for $0 < |z - z_0| < \delta$. So by the Theorem on Classification of Zeros, there exists an $\widetilde{m} \in \mathbb{N}$, which is the order of z_0 as a zero of f and a $\widetilde{g}$ holomorphic in D such that $\widetilde{g}(z_0) \neq 0$. Then for $|z - z_0| < R$, we have $(z - z_0)^{\widetilde{m}} \widetilde{g}(z) = (z - z_0)^m g(z)$. We show that this implies $\widetilde{m} = m$. For if $\widetilde{m} > m$, then we get the contradiction that

$$0 \neq g(z_0) = \lim_{z \to z_0} g(z) = \lim_{z \to z_0} (z - z_0)^{\widetilde{m}-m} \widetilde{g}(z) = 0 \cdot \widetilde{g}(z_0) = 0,$$

while if $m > \widetilde{m}$, then we get the contradiction that

$$0 \neq \widetilde{g}(z_0) = \lim_{z \to z_0} \widetilde{g}(z) = \lim_{z \to z_0} (z - z_0)^{m-\widetilde{m}} g(z) = 0 \cdot g(z_0) = 0.$$

Consequently, $m = \widetilde{m}$, and so z_0 is a zero of order $m = \widetilde{m}$.

Solution to Exercise 4.18

(1) We have

$$f(z) = (1 + z^2)^4 = ((z - i)(z + i))^4 = (z - i)^4 (z + i)^4,$$

and so with $g(z) := (z + i)^4$, g is entire, $g(i) = (2i)^4 = 16 \neq 0$ and $f(z) = (z - i)^4 g(z)$. So i is a zero of f of order 4.

(2) We have

$$f(2n\pi i) = 1 - 1 = 0, \text{ and}$$

$$f'(2n\pi i) = \exp z \Big|_{z=2n\pi i} = 1 \neq 0.$$

So $2n\pi i$ is a zero of f of order 1.

(3) $f(0) = 1 - 1 + \dfrac{1}{2}(0)^2 = 0$, and we have

$$f(z) = \cos z - 1 + \frac{1}{2}(\sin z)^2 = \cos z - 1 + \frac{1}{2} \cdot \frac{(1 - \cos(2z))}{2}$$

$$= \cos z - \frac{3}{4} - \frac{1}{4}\cos(2z)$$

$$= \left(1 - \frac{z^2}{2!} + \frac{z^4}{4!} - \frac{z^6}{6!} + - \cdots\right) - \frac{3}{4}$$

$$\qquad - \frac{1}{4}\left(1 - \frac{4z^2}{2!} + \frac{16z^4}{4!} - \frac{2^6 z^6}{6!} + - \cdots\right)$$

$$= \underbrace{\left(1 - \frac{3}{4} - \frac{1}{4}\right)}_{0} + \underbrace{\left(-\frac{1}{2!} + \frac{1}{4} \cdot \frac{4}{2!}\right)}_{0} z^2 + \underbrace{\left(\frac{1}{4!} - \frac{1}{4} \cdot \frac{16}{4!}\right)}_{\neq 0} z^4 + \cdots,$$

and so z_0 is a zero of order 4.

Solution to Exercise 4.19

First we note that if z is a point that is distinct from z_0 in the disc, then $f(z) \neq 0$. By the result on the classification of zeros, $f(z) = (z - z_0)g(z)$, where g is holomorphic in the disc, and $g(z_0) \neq 0$. Thus

$$\frac{1}{2\pi i}\int_\gamma \frac{zf'(z)}{f(z)}dz$$

$$= \frac{1}{2\pi i}\int_\gamma \frac{z(1 \cdot g(z) + (z - z_0) \cdot g'(z))}{(z - z_0)g(z)}dz$$

$$= \frac{1}{2\pi i}\int_\gamma \frac{\dfrac{z(g(z) + (z - z_0) \cdot g'(z))}{g(z)}}{(z - z_0)}dz$$

$$= \left.\frac{z(g(z) + (z - z_0) \cdot g'(z))}{g(z)}\right|_{z=z_0} \quad \text{(Cauchy Integral Formula)}$$

$$= \frac{z_0(g(z_0) + 0 \cdot g'(z_0))}{g(z_0)}$$

$$= z_0.$$

Solution to Exercise 4.20

By the result on the classification of zeros, $f(z) = (z - z_0)^m g(z)$, where g is holomorphic in D and $g(z_0) \neq 0$. Thus

$$(f(z))^2 = (z - z_0)^{2m} \underbrace{(g(z))^2}_{=:G(z)}.$$

Clearly $(f(z_0))^2 = 0$, and G is holomorphic in D with $G(z_0) = (g(z_0))^2 \neq 0$. Hence z_0 is a zero of $z \mapsto (f(z))^2$ of order $2m$. Also,

$$\begin{aligned}
f'(z) &= m(z - z_0)^{m-1} g(z) + (z - z_0)^m g'(z) \\
&= (z - z_0)^{m-1} \underbrace{(mg(z) + (z - z_0)g'(z))}_{=:g_1(z)},
\end{aligned}$$

and so $f'(z_0) = (z_0 - z_0)^{m-1} g_1(z_0) \overset{(m>1)}{=} 0 \cdot g_1(z_0) = 0$. As g_1 is holomorphic and

$$g_1(z_0) = mg(z_0) + 0 \cdot g'(z_0) = mg(z_0) + 0 = mg(z_0) \neq 0,$$

it follows that z_0 is a zero of f' of order $m - 1$.

Solution to Exercise 4.21

Consider the function $f : \mathbb{C} \to \mathbb{C}$ given by

$$f(x, y) = x \sin \frac{1}{x} \text{ if } x \neq 0,$$

and $f(0, *) = 0$. Then f is obviously continuous at any point (x_0, y_0) where $x_0 \neq 0$. Moreover, since

$$|f(x, y_0) - f(0, y_0)| = \left| x \sin \frac{1}{x} - 0 \right| = |x| \left| \sin \frac{1}{x} \right| \leq |x| \cdot 1 = |x - 0|$$

for all $x \neq 0$. Thus f is continuous also at the points $(0, *)$. Hence f is continuous everywhere in $\mathbb{C}$. By the definition of f, 0 is a zero of f. 0 is clearly not an isolated zero of f, since

$$f\left(\frac{1}{n\pi}, 0\right) = \frac{1}{n\pi} \sin(n\pi) = 0, \quad n \in \mathbb{N}.$$

Also, f is not identically zero in any disc centered at 0 because for $n \in \mathbb{N}$,

$$f\left(\frac{2}{(2n+1)\pi}, 0\right) = \frac{2}{(2n+1)\pi} \sin\left((2n+1)\frac{\pi}{2}\right) = \frac{2}{(2n+1)\pi}(-1)^n \neq 0.$$

Solution to Exercise 4.22

We know that for all $x_1, x_2 \in \mathbb{R}$,

$$\cos(x_1 + x_2) = (\cos x_1)(\cos x_2) - (\sin x_1)(\sin x_2). \qquad (5.22)$$

Fix $x \in \mathbb{R}$ and consider the entire function f given by

$$f(z) := \cos(z + x) - \Big((\cos z)(\cos x) - (\sin z)(\sin x)\Big), \quad z \in \mathbb{C}.$$

We have $f(y) = 0$ for all $y \in \mathbb{R}$, thanks to (5.22), and so by the Identity Theorem, $f(z) = 0$ for all $z \in \mathbb{C}$, that is,

$$\cos(z + x) = (\cos z)(\cos x) - (\sin z)(\sin x), \quad z \in \mathbb{C}. \qquad (5.23)$$

But the choice of $x \in \mathbb{R}$ was arbitrary, and so (5.23) holds for *all* $x \in \mathbb{R}$. Next, fix a $z \in \mathbb{C}$. Consider the entire function g given by

$$g(w) := \cos(z + w) - \Big((\cos z)(\cos w) - (\sin z)(\sin w)\Big), \quad w \in \mathbb{C}.$$

Then we have $g(x) = 0$ for all $x \in \mathbb{R}$ by (5.23). Another application of the Identity Theorem yields $g(w) = 0$ for all $w \in \mathbb{C}$. Hence

$$\cos(z + w) = (\cos z)(\cos w) - (\sin z)(\sin w), \quad w \in \mathbb{C}. \qquad (5.24)$$

But the choice of $z \in \mathbb{C}$ was arbitrary. Consequently, (5.24) holds for all $z \in \mathbb{C}$ (and for all $w \in \mathbb{C}$).

Solution to Exercise 4.23

Suppose $f, g \in \mathrm{Hol}(D)$ are such that

$$(f \cdot g)(z) = f(z) \cdot g(z) = 0, \quad z \in D. \qquad (5.25)$$

Suppose there exists $z_0 \in D$ such that $f(z_0) \neq 0$. By the continuity of f, there exists a $\delta > 0$ such that $f(z) \neq 0$ whenever $|z - z_0| < \delta$. (5.25) then implies $g(z) = 0$ for $|z - z_0| < \delta$. By the Identity Theorem, $g \equiv 0$ in D. So $\mathrm{Hol}(D)$ has no zero divisors.

On the other hand, $C(D)$ is not an integral domain, and we show this below. Let $z_0 \in D$ and let $\delta > 0$ be such that the disc

$$\Delta := \{z \in D : |z - z_0| < \delta\} \subset D.$$

Consider the continuous function $\varphi : \mathbb{R} \to \mathbb{R}$ defined by

$$\varphi(t) = \begin{cases} 0 & \text{if } t \leq 0, \\ t & \text{if } t > 0. \end{cases}$$

Define f, g by

$$f(z) := \varphi(\mathrm{Re}(z - z_0)),$$
$$g(z) := \varphi(-\mathrm{Re}(z - z_0)),$$

for $z \in D$.

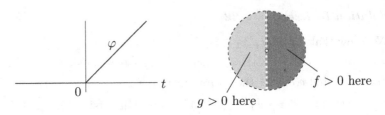

Fig. 5.21   Construction of $f, g \in C(D)$ using φ.

Being the composition of continuous functions, $f, g \in C(D)$. Also $f(z) > 0$ for all z in the right half of Δ, and so $f \neq 0$ in $C(D)$. Similarly $g(z) > 0$ for all z belonging to the left half of Δ, and so $g \neq 0$ in $C(D)$. Nevertheless, $f \cdot g = 0$.

Solution to Exercise 4.24

(1) False. Take $D = \mathbb{C}$, $f = \exp$, $g = 1$. Then for all $n \in \mathbb{N}$, we have $f(2\pi i n) = \exp(2\pi i n) = 1 = g(2\pi i n)$, but $f \neq g$ (for example, because $f(i\pi) = -1 \neq 1 = g(i\pi)$).

(2) True.

(3) True. Let $\gamma(t) = x(t) + iy(t)$, $t \in [a, b]$. Consider a point t_0 where either $x'(t_0)$ or $y'(t_0)$ is nonzero. (If they are both always 0, then $a = b$, a contradiction.) Suppose $x'(t_0) > 0$ (the other cases are handled similarly). Then $x(t) > 0$ in a neighbourhood of t_0. So x is increasing there. Take $t_n = t_0 + \frac{1}{n}$, $n \geq N$, where N is large enough so that $t_0 + \frac{1}{N} \in [a, b]$. Set $z_n = \gamma(t_n)$. Then $(z_n)_{n \geq N}$ is a sequence of *distinct* points (since their real parts ate distinct), which converges to $\gamma(t_0)$. By the Identity Theorem, $f = g$ in D.

(4) True. By the Taylor expansion around w, $f = g$ in a disc with center w, and so by the Identity Theorem, $f = g$ in D.

Solution to Exercise 4.25

Let $K = \{z \in \mathbb{C} : |z| \leq 1\}$. For each $z \in K$, there exists a smallest $n(z) \in \{0, 1, 2, 3, \cdots\}$ such that for all w near z,

$$f(w) = \sum_{n=0}^{\infty} c_n(z)(w - z)^n$$

and $c_{n(z)}(z) = 0$. Hence $(f^{(n(z))}(z))/((n(z))!) = 0$, and so $f^{(n(z))}(z) = 0$. Let $\varphi : K \to \mathbb{N} \cup \{0\}$ be defined by $\varphi(z) = n(z)$. Since K is uncountable, while $\mathbb{N} \cup \{0\}$ is countable, there exists an N such that $\varphi^{-1}(N)$ is infinite. Let $(z_n)_{n \in \mathbb{N}}$ be a sequence of distinct points in $\varphi^{-1}(N)$. As K is compact, this has a convergent subsequence $(z_{n_k})_{k \in \mathbb{N}}$ with limit, say $z_* \in K$. As $f^{(N)}(z_{n_k}) = 0$ for all k, by the Identity Theorem (applied to $f^{(N)}$), we have $f^{(N)} = 0$ in K, and so also in $\mathbb{C}$. By Taylor's Theorem,

$$f(z) = \sum_{n=0}^{\infty} \frac{f^{(n)}(0)}{n!} z^n = \sum_{n=0}^{N-1} \frac{f^{(n)}(0)}{n!} z^n,$$

for all $z \in \mathbb{C}$, and so f is a polynomial.

Solution to Exercise 4.26

Let $z_0 \in D$ be such that $|f(z_0)| \geq |f(z)|$ for all $z \in D$. By the Maximum Modulus Theorem, f is constant in D, a contradiction.

Solution to Exercise 4.27

Let $f(z_0) \neq 0$. Then $|f(z_0)| > 0$ and so for all $z \in D$, $|f(z)| \geq |f(z_0)| > 0$, implying that for all $z \in D$, $f(z) \neq 0$. Now consider the holomorphic function $g := 1/f$ in D. We have

$$|g(z_0)| = \frac{1}{|f(z_0)|} \geq \frac{1}{|f(z)|} = |g(z)|, \quad z \in D,$$

and so by the Maximum Modulus Theorem, g is constant. But then f is constant too.

Solution to Exercise 4.28

Let z_0 be a maximizer, which exists since $z \mapsto |f(z)|$ is continuous and $K := \{z \in \mathbb{C} : |z| \leq 1\}$ is compact. But z_0 can't be in the interior of L: indeed, if $|z_0| < 1$, then by the Maximum Modulus Theorem (applied to f on $\mathbb{D} := \{z \in \mathbb{C} : |z| < 1\}$), f would be constant in $\mathbb{D}$, which it clearly isn't. Hence $z \in \mathbb{T} := \{z \in \mathbb{C} : |z| = 1\}$. So

$$\max_{z \in K} |f(z)| = \max_{|z|=1} |f(z)| = \max_{t \in [0, 2\pi)} |\exp(2it) - 2| = |-1 - 2| = 3.$$

Similarly, if z_1 is a minimizer, z_1 can't be in the interior of K. Indeed, $z_1^2 - 2 \neq 0$, and so by the Minimum Modulus Theorem, f would be a constant, a contradiction. So $z_1 \in \mathbb{T}$. Hence

$$\min_{z \in K} |f(z)| = \min_{|z|=1} |f(z)| = \min_{t \in [0, 2\pi)} |\exp(2it) - 2| = |1 - 2| = 1.$$

See Figure 5.22.

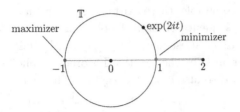

Fig. 5.22   Maximizer and minimizer for $|z^2 - 2|$ in the unit disc.

Solution to Exercise 4.29

For $z \in \mathbb{A}_1 := \{z \in \mathbb{C} : 0 < |z - 1| < 1\}$, we have

$$
\begin{aligned}
\frac{1}{z(z-1)} &= \frac{1}{(z-1+1)(z-1)} \\
&= \frac{1}{z-1}(1 - (z-1) + (z-1)^2 - (z-1)^3 + -\cdots) \\
&= \frac{1}{z-1} - 1 + (z-1) - (z-1)^2 + (z-1)^3 - +\cdots.
\end{aligned}
$$

On the other hand, for $z \in \widetilde{\mathbb{A}}_1 := \{z \in \mathbb{C} : 1 < |z-1|\}$

$$
\begin{aligned}
\frac{1}{z(z-1)} &= \frac{1}{(z-1+1)(z-1)} = \frac{1}{(z-1)^2\left(1 + \dfrac{1}{z-1}\right)} \\
&= \frac{1}{(z-1)^2}\left(1 - \frac{1}{z-1} + \frac{1}{(z-1)^2} - \frac{1}{(z-1)^3} - +\cdots\right) \\
&= \frac{1}{(z-1)^2} - \frac{1}{(z-1)^3} + \frac{1}{(z-1)^4} - \frac{1}{(z-1)^5} + -\cdots.
\end{aligned}
$$

Solution to Exercise 4.30

By the result on classification of zeros, $f(z) = (z - z_0)^m g(z)$, $z \in D$, where g is holomorphic in D and $g(z_0) \neq 0$. Also, since z_0 is the only zero of f in D, $g(z) \neq 0$ for all $z \in D$. So $1/g$ is holomorphic, and it has a Taylor expansion in a disc with center z_0: there exists an $R > 0$ such that

$$
\frac{1}{g(z)} = \sum_{n=0}^{\infty} c_n(z - z_0)^n \text{ for } |z - z_0| < R,
$$

and $c_0 \neq 0$ (because $g(z_0) \neq 0$). Thus for $0 < |z - z_0| < R$,

$$
\frac{1}{f(z)} = \frac{1}{(z-z_0)^m g(z)} = \frac{1}{(z-z_0)^m} \sum_{n=0}^{\infty} c_n (z-z_0)^n
$$

$$
= \frac{c_0}{(z-z_0)^m} + \frac{c_1}{(z-z_0)^{m-1}} + \cdots + \frac{c_{m-1}}{z-z_0} + \sum_{n=0}^{\infty} c_{m+n}(z-z_0)^n.
$$

Hence $1/f$ has a pole of order m at z_0.

Solution to Exercise 4.31

$z \mapsto (z-z_0)^m f(z)$ has a holomorphic extension, say h, to D. Also,

$$
\neg \left(\lim_{z \to z_0} (z-z_0)^m f(z) = 0 \right),
$$

and so $h(z_0) \neq 0$. Moreover, since $f(z) \neq 0$ for $z \in D$, also $h(z) \neq 0$ for all $z \in D$. Hence

$$
\frac{1}{f(z)} = \frac{(z-z_0)^m}{h(z)} \quad \text{for all } z \in D \setminus \{z_0\}
$$

and g defined by

$$
g(z) := \frac{(z-z_0)^m}{h(z)}, \quad z \in D
$$

is holomorphic in D. Since $\dfrac{1}{h(z_0)} \neq 0$, z_0 is a zero of g of order m.

Solution to Exercise 4.32

We must have $c_n = 0$ for all $n < -m$. Thus

$$
f(z) = \frac{c_{-m}}{(z-z_0)^m} + \frac{c_{-m+1}}{(z-z_0)^{m-1}} + \cdots + \frac{c_{-1}}{z-z_0} + \sum_{n=0}^{\infty} c_n (z-z_0)^n.
$$

So $(z-z_0)^m f(z) = c_{-m} + c_{-m+1}(z-z_0) + \cdots + c_{-1}(z-z_0)^{m-1} + \cdots$. Hence $(z-z_0)^m f(z)$ has a holomorphic extension, say g, to

$$
\Delta := \{z \in \mathbb{C} : |z - z_0| < R\}.
$$

So for $|z - z_0| < R$, $g(z) = c_{-m} + c_{-m+1}(z-z_0) + \cdots + c_{-1}(z-z_0)^{m-1} + \cdots$. By Taylor's Theorem,

$$
c_{-1} = \frac{1}{(m-1)!} \frac{d^{m-1} g}{dz^{m-1}}(z_0).
$$

But $g^{(m-1)}$ is holomorphic in Δ and in particular, continuous at z_0. So

$$g^{(m-1)}(z_0) = \lim_{z \to z_0} g^{(m-1)}(z).$$

Also, for $0 < |z - z_0| < R$, $g(z) = (z - z_0)^m f(z)$, and so for $z \neq z_0$ in Δ,

$$g^{(m-1)}(z) = \frac{d^{m-1}}{dz^{m-1}}((z - z_0)^m f(z)).$$

Hence

$$c_{-1} = \frac{1}{(m-1)!} g^{(m-1)}(z_0) = \frac{1}{(m-1)!} \lim_{z \to z_0} g^{(m-1)}(z)$$

$$= \frac{1}{(m-1)!} \lim_{z \to z_0} \frac{d^{m-1}}{dz^{m-1}}((z - z_0)^m f(z)).$$

Solution to Exercise 4.33

(1) True, since $c_{-1} = 1 \neq 0$, and $c_{-2} = c_{-3} = \cdots = 0$.
(2) True.
(3) True.
(4) True.
(5) True.

Solution to Exercise 4.34

(1) $\sin z$ does not have a singularity at 0, and for $z \in \mathbb{C}$,

$$\sin z = z - \frac{z^3}{3!} + \frac{z^5}{5!} - + \cdots.$$

(2) $\sin \dfrac{1}{z}$ has an essential singularity at 0, since for $z \neq 0$,

$$\sin \frac{1}{z} = \cdots - + \frac{1}{5! z^5} - \frac{1}{3! z^3} + \frac{1}{z}.$$

(3) $\dfrac{\sin z}{z}$ has a removable singularity at 0, since we have that

$$\lim_{z \to 0} z \cdot \frac{\sin z}{z} = \lim_{z \to 0} \sin z = 0.$$

Also, $\dfrac{\sin z}{z} = 1 - \dfrac{1}{3!} z^2 + \dfrac{1}{5!} z^4 - \dfrac{1}{7!} z^6 + - \cdots$ for $z \neq 0$.

(4) $\dfrac{\sin z}{z^2}$ has a pole of order 1 at 0, since for $z \neq 0$,

$$\frac{\sin z}{z^2} = \frac{1}{z} - \frac{z}{3!} + \frac{z^3}{5!} - \frac{z^5}{7!} + - \cdots .$$

(5) $1/(\sin(1/z))$ does not have an isolated singularity at 0 because with $z_n = 1/(n\pi)$, $n \in \mathbb{N}$, we have

$$\sin \frac{1}{z_n} = \sin(n\pi) = 0$$

and $z_n = \dfrac{1}{n\pi} \overset{n \to \infty}{\longrightarrow} 0$. (We had also observed this in Example 4.13.)

(6) $z \sin \dfrac{1}{z}$ has an essential singularity at 0 since for $z \neq 0$

$$z \sin \frac{1}{z} = \cdots - \frac{1}{5! z^4} - \frac{1}{3! z^2} + 1.$$

Solution to Exercise 4.35

(1) False. $\lim\limits_{x \nearrow 0} |e^{\frac{1}{x}}| = \lim\limits_{x \nearrow 0} e^{\frac{1}{x}} = 0$, and so $\neg \left(\lim\limits_{z \to 0} \left| \exp \dfrac{1}{z} \right| = +\infty \right)$.

(2) True. There exists an $R > 0$ such that

$$f(z) = \frac{c_{-m}}{(z - z_0)^m} + \frac{c_{-m+1}}{(z - z_0)^{m-1}} + \cdots + \frac{c_{-1}}{z - z_0} + \sum_{n=0}^{\infty} c_n (z - z_0)^n,$$

for $0 < |z - z_0| < R$, and so with

$$p := c_{-m} + c_{-m+1}(z - z_0) + \cdots + c_{-1}(z - z_0)^{m-1},$$

we have for $0 < |z - z_0| < R$,

$$f(z) - \frac{p(z)}{(z - z_0)^m} = \sum_{n=0}^{\infty} c_n (z - z_0)^n.$$

(3) True. Let the order of 0 as a zero of f be m. (Take $m = 0$ if $f(0) \neq 0$.) Then there exists a holomorphic function g such that $f(z) = z^m g(z)$ and $g(0) \neq 0$. Hence for $n > m$, and $z \neq 0$,

$$\frac{f(z)}{z^n} = \frac{z^m g(z)}{z^n} = \frac{g(z)}{z^{n-m}}.$$

Thus

$$\lim_{z \to 0} \left| \frac{f(z)}{z^n} \right| = \lim_{z \to 0} \frac{|g(z)|}{|z|^{n-m}} = |g(z_0)| \cdot \lim_{z \to 0} \frac{1}{|z|^{n-m}} = +\infty$$

since $g(z_0) \neq 0$ and $n > m$.

(4) True. In some punctured disc $D = \{z \in \mathbb{C} : 0 < |z - z_0| < R\}$, f, g are nonzero, and there exist holomorphic functions h_f, h_g such that $h_f(z_0) \neq 0$, $h_g(z_0) \neq 0$, and for all $z \in D$,

$$\frac{1}{f(z)} = (z - z_0)^{m_f} h_f(z), \quad \frac{1}{g(z)} = (z - z_0)^{m_g} h_g(z).$$

Thus $h_f(z_0) h_g(z_0) \neq 0$ and for all $z \in D$

$$\frac{1}{f(z) g(z)} = (z - z_0)^{m_f + m_g} h_f(z) h_g(z).$$

Consequently, fg has a pole of order $m_f + m_g$ at z_0.

Solution to Exercise 4.36

Consider f given by

$$f(z) = \left(\exp \frac{1}{z} \right) + \exp \left(\frac{1}{1 - z} \right), \quad z \in \mathbb{C} \setminus \{0, 1\}.$$

Then f is holomorphic in $\mathbb{C} \setminus \{0, 1\}$. The function $\exp(1/(1 - z))$ is holomorphic in a neighbourhood of $z = 0$, while the function $\exp(1/z)$ has an essential singularity at 0. Thus, their sum, namely f, has an essential singularity at 0. (Why?) On the other hand $\exp(1/z)$ is holomorphic in a neighbourhood of 1, while $\exp(1/(1 - z))$ has an essential singularity there. So f has an essential singularity at $z = 1$.

Solution to Exercise 4.37

We have seen that if z_0 is an isolated singularity of a function g with the Laurent series expansion

$$g(z) = \sum_{n \in \mathbb{Z}} c_n (z - z_0)^n$$

for $0 < |z - z_0| < R$ and for some $R > 0$, and there are infinitely many indices $n < 0$ such that $c_n \neq 0$, then z_0 is an essential singularity of g.

However, for the given f, the annulus for the Laurent expansion

$$z^{-1} + z^{-2} + z^{-3} + \cdots$$

is given by $|z| > 1$. The correct annulus to consider for deciding the nature of the singularity at $z = 0$ is of the form $0 < |z| < R$ for some $R > 0$. In fact, for $|z| < 1$ we have

$$f(z) = -\frac{1}{1 - z} = -(1 + z + z^2 + z^3 + \cdots),$$

showing that f is holomorphic for $|z| < 1$, and f does not have a singularity at $z = 0$.

Solution to Exercise 4.38

It is clear that z_0 is an isolated singularity of fg. Indeed, since f and g both have an isolated singularity at z_0, we have that f is holomorphic in a punctured disc $0 < |z - z_0| < R_f$ for some $R_f > 0$, and g is holomorphic in a punctured disc $0 < |z - z_0| < R_g$ for some $R_g > 0$. Thus fg is holomorphic in the punctured disc $0 < |z - z_0| < \min\{R_f, R_g\}$.

Suppose that fg has a removable singularity or a pole at z_0. Then there exists an $m \geq 1$ such that

$$\lim_{z \to z_0} (z - z_0)^m f(z)g(z) = 0.$$

Since f has a pole at z_0, say of order m_f, f is nonzero near z_0 and

$$f(z) = \frac{c_{-m_f}}{(z - z_0)^{m_f}} + \frac{c_{-m_f+1}}{(z - z_0)^{m_f-1}} + \cdots + \frac{c_{-1}}{z - z_0} + \sum_{n=0}^{\infty} c_n(z - z_0)^n,$$

for $0 < |z - z_0| < R$ and for some $R > 0$. Here $c_{-m_f} \neq 0$. So for $z \neq z_0$, but near z_0, we have

$$(z - z_0)^m g(z) = \frac{1}{(z - z_0)^{m_f} f(z)} \cdot \underbrace{(z - z_0)^{m_f}}_{\to 0} \underbrace{(z - z_0)^m f(z)g(z)}_{\to 0}$$

$$\xrightarrow{z \to z_0} \frac{1}{c_{-m_f}} \cdot 0 \cdot 0 = 0.$$

So g must have a pole at z_0 or a removable singularity at z_0, a contradiction. Consequently fg has an essential singularity at z_0.

Solution to Exercise 4.39

Set $\epsilon := 1/n =: \delta \ (> 0)$. By the Casorati-Weierstrass Theorem, there exists a z_n in the punctured disc around z_0 with radius δ such that $|f(z_n) - w| < \epsilon$. That is, $|z_n - z_0| < 1/n$ and $|f(z_n) - w| < \epsilon$. Hence $(z_n)_{n \in \mathbb{N}}$ converges to z_0, and $(f(z_n))_{n \in \mathbb{N}}$ converges to w.

Solution to Exercise 4.40

$1 + \exp z = 0$ if and only if $z \in \{\pi i + 2\pi n i : n \in \mathbb{Z}\}$. So

$$f(z) := \frac{\text{Log}(z)}{1 + \exp z}$$

is holomorphic in $(\mathbb{C} \setminus (-\infty, 0]) \setminus \{\pi i + 2\pi n i : n \in \mathbb{Z}\}$. f has poles of order 1 at the points $\{\pi i + 2\pi n i : n \in \mathbb{Z}\}$, of which two lie inside the given path γ: $-\pi$ and $3\pi i$. See Figure 5.23.

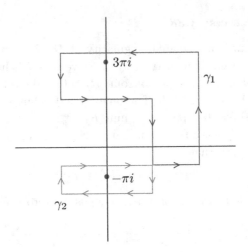

Fig. 5.23   The curves γ_1 and γ_2.

We have
$$\int_\gamma f(z)dz = \int_{\gamma_1} f(z)dz + \int_{\gamma_2} f(z)dz = 2\pi i(\mathrm{res}(f, 3\pi i) - \mathrm{res}(f, -\pi i)).$$
So we need to calculate $\mathrm{res}(f, 3\pi i)$ and $\mathrm{res}(f, -\pi i)$. We can write
$$\frac{\mathrm{Log}(z)}{1 + \exp z} = \frac{c_{-1,3\pi i}}{z - 3\pi i} + h_{3\pi i},$$
where $h_{3\pi i}$ is holomorphic in a neighbourhood of $3\pi i$. Thus
$$c_{-1,3\pi i} = \lim_{z \to 3\pi i} \frac{(z - 3\pi i)\mathrm{Log}(z)}{1 + \exp z} = \lim_{z \to 3\pi i} \frac{z - 3\pi i}{\exp z - \exp(3\pi i)} \cdot \mathrm{Log}(z)$$
$$= \frac{1}{\exp z|_{z = 3\pi i}} \cdot \mathrm{Log}(3\pi i) = -1\Big(\log|3\pi i| + i\frac{\pi}{2}\Big)$$
$$= -\log 3 - \log \pi - i\frac{\pi}{2}.$$
We can write
$$\frac{\mathrm{Log}(z)}{1 + \exp z} = \frac{c_{-1,-\pi i}}{z - (-\pi i)} + h_{-\pi i},$$
where $h_{-\pi i}$ is holomorphic in a neighbourhood of $-\pi i$. Thus
$$c_{-1,-\pi i} = \lim_{z \to -\pi i} \frac{(z - (-\pi i))\mathrm{Log}(z)}{1 + \exp z} = \lim_{z \to -\pi i} \frac{z - (-\pi i)}{\exp z - \exp(-\pi i)} \cdot \mathrm{Log}(z)$$
$$= \frac{1}{\exp z|_{z = -\pi i}} \cdot \mathrm{Log}(-\pi i) = -1\Big(\log|-\pi i| + i\Big(-\frac{\pi}{2}\Big)\Big)$$
$$= -\log \pi + i\frac{\pi}{2}.$$
So $\displaystyle\int_\gamma \frac{\mathrm{Log}(z)}{1 + \exp z}dz = 2\pi i\Big(-\log 3 - \log \pi - i\frac{\pi}{2} + \log \pi - i\frac{\pi}{2}\Big) = 2\pi^2 - (2\pi \log 3)i.$

Solution to Exercise 4.41

Let γ be the circular path given by $\gamma(\theta) = \exp(i\theta)$ for $\theta \in [0, 2\pi)$. Then

$$\int_0^{2\pi} \frac{\cos\theta}{5 + 4\cos\theta} d\theta = \int_\gamma \frac{\frac{z + \frac{1}{z}}{2}}{5 + 4\frac{z + \frac{1}{z}}{2}} \cdot \frac{1}{iz} dz = \int_\gamma \frac{z^2 + 1}{2iz(2z^2 + 5z + 2)} dz$$

$$= \int_\gamma \frac{z^2 + 1}{2iz(2z + 1)(z + 2)} dz.$$

Let f be defined by

$$f(z) := \frac{z^2 + 1}{2iz(2z + 1)(z + 2)}.$$

Then f has three poles, at 0, $-1/2$, -2, and each is of order 1. Of these, the poles at 0 and $-1/2$ lie inside γ. So by the Residue Theorem,

$$\int_0^{2\pi} \frac{\cos\theta}{5 + 4\cos\theta} d\theta$$

$$= 2\pi i \Big(\text{res}(f, 0) + \text{res}(f, -1/2) \Big)$$

$$= 2\pi i \left(\lim_{z \to 0} \frac{z \cdot (z^2 + 1)}{2iz(2z + 1)(z + 2)} + \lim_{z \to -1/2} \frac{(z + 1/2) \cdot (z^2 + 1)}{2iz(2z + 1)(z + 2)} \right)$$

$$= 2\pi i \cdot \left(\frac{1}{2i \cdot 1 \cdot 2} + \frac{1 \cdot \frac{5}{4}}{2i \cdot \left(-\frac{1}{2}\right) \cdot 2 \cdot \frac{3}{2}} \right) = 2\pi i \cdot \left(\frac{1}{4i} - \frac{5}{12i} \right)$$

$$= -\frac{\pi}{3}.$$

Solution to Exercise 4.42

(1) Let f_1 be defined by

$$f_1(z) = \frac{1}{1 + z^2}.$$

Then f_1 has poles at i and $-i$, both of order 1. Hence

$$\int_0^\infty \frac{1}{1 + x^2} dx = \frac{1}{2} \cdot 2\pi i \cdot \text{res}(f_1, i) = \pi i \cdot \lim_{z \to i} \frac{z - i}{1 + z^2}$$

$$= \pi i \cdot \lim_{z \to i} \frac{1}{z + i} = \pi i \cdot \frac{1}{2i} = \frac{\pi}{2}.$$

(2) Let f_2 be defined by

$$f_2(z) = \frac{1}{(a^2 + z^2)(b^2 + z^2)}.$$

Then f_2 has poles at ai, $-ai$, bi, $-bi$, all of order 1. As f_2 is even,

$$\int_0^\infty \frac{1}{(a^2+x^2)(b^2+x^2)}\, dx = \frac{1}{2}\cdot 2\pi i\Big(\operatorname{res}(f_2, ai) + \operatorname{res}(f_2, bi)\Big)$$

$$= \pi i\left(\frac{1}{(b^2-a^2)2ai} + \frac{1}{(a^2-b^2)2bi}\right)$$

$$= \frac{\pi}{2(a^2-b^2)}\left(\frac{1}{b}-\frac{1}{a}\right) = \frac{\pi}{2ab(a+b)}.$$

(3) Let f_3 be defined by

$$f_3(z) = \frac{1}{(1+z^2)^2}.$$

Then f_3 has poles at i and $-i$, both of order 2. We have

$$\int_0^\infty \frac{1}{(1+x^2)^2}\, dx = \frac{1}{2}\cdot 2\pi i\cdot \operatorname{res}(f_3, i)$$

$$= \frac{\pi i}{1!}\cdot \lim_{z\to i}\frac{d}{dz}\left((z-i)^2\cdot \frac{1}{(z-i)^2(z+i)^2}\right)$$

$$= \pi i\cdot \lim_{z\to i}\frac{-2}{(z+i)^3} = \pi i\cdot \frac{-2}{-8i} = \frac{\pi}{4}.$$

(4) Let f_4 be defined by

$$f_4(z) = \frac{1+z^2}{1+z^4}.$$

Then f_4 has poles at

$$p_1 = \exp\left(\frac{\pi i}{4}\right),\ p_2 = \exp\left(\frac{3\pi i}{4}\right),\ p_3 = \exp\left(\frac{5\pi i}{4}\right),\ p_4 = \exp\left(\frac{7\pi i}{4}\right),$$

all of order 1. Hence

$$\int_0^\infty \frac{1+x^2}{1+x^4}\, dx$$

$$= \frac{1}{2}\cdot 2\pi i\Big(\operatorname{res}(f_4, p_1) + \operatorname{res}(f_4, p_2)\Big) = \pi i\left(\frac{1+p_1^2}{4p_1^3} + \frac{1+p_2^2}{4p_2^3}\right)$$

$$= \pi i\left(\frac{p_1}{4p_1^4} + \frac{1}{4p_1} + \frac{p_2}{4p_2^4} + \frac{1}{4p_2}\right) = \pi i\left(-\frac{p_1+p_2}{4} + \frac{1}{4p_1} + \frac{1}{4p_2}\right)$$

$$= \pi i\left(-\frac{\exp(i\pi/4)-\exp(-\pi i/4)}{4} + \frac{\exp(-i\pi/4)-\exp(i\pi/4)}{4}\right)$$

$$= \pi i\left(-\frac{i\sin(\pi/4)}{2} + \frac{-i\sin(\pi/4)}{2}\right) = \pi i\cdot(-i)\cdot\frac{1}{\sqrt{2}} = \frac{\pi}{\sqrt{2}}.$$

Solution to Exercise 4.43

By the Residue Theorem,

$$
\int_C \frac{\exp z}{z^{n+1}} dz = 2\pi i \cdot \operatorname{res}\left(\frac{\exp z}{z^{n+1}}, 0\right) = \frac{2\pi i}{n!} \cdot \lim_{z\to 0} \frac{d^n}{dz^n}\left(z^{n+1} \cdot \frac{\exp z}{z^{n+1}}\right)
$$

$$
= \frac{2\pi i}{n!} \cdot \lim_{z\to 0} \frac{d^n}{dz^n} \exp z = \frac{2\pi i}{n!} \cdot \lim_{z\to 0} \exp z = \frac{2\pi i}{n!} \cdot \exp 0
$$

$$
= \frac{2\pi i}{n!} \cdot 1 = \frac{2\pi i}{n!}.
$$

Hence

$$
\frac{2\pi i}{n!} = \int_0^{2\pi} \frac{\exp(\cos\theta + i\sin\theta)}{\cos((n+1)\theta) + i\sin((n+1)\theta)} \cdot i(\cos\theta + i\sin\theta) d\theta
$$

$$
= i \int_0^{2\pi} \exp(\cos\theta + i\sin\theta) \cdot \Big(\cos(n\theta) - i\sin(n\theta)\Big) d\theta
$$

$$
= i \int_0^{2\pi} \exp(\cos\theta) \Big(\cos(n\theta - \sin\theta) - i\sin(n\theta - \sin\theta)\Big) d\theta.
$$

Equating the imaginary parts, we obtain

$$
\int_0^{2\pi} \exp(\cos\theta) \cdot \cos(n\theta - \sin\theta) d\theta = \frac{2\pi}{n!}.
$$

Solution to Exercise 4.44

For z in a small punctured disc D centered at z_0, $f(z) \neq 0$ and

$$
f(z) = (z - z_0)h(z) \tag{5.26}
$$

for some holomorphic function h such that $h(z_0) \neq 0$. From (5.26) we have $f'(z) = h(z) + (z - z_0)h'(z)$ and in particular, $f'(z_0) = h(z_0)$. Now

$$
\frac{1}{f(z)} = \frac{1}{(z - z_0)h(z)} \quad \text{for } z \in D \setminus \{z_0\},
$$

and since $\dfrac{1}{h}$ is holomorphic in D,

$$
\frac{1}{h(z)} = d_0 + d_1(z - z_0) + \cdots \quad \text{for } z \in D,
$$

where $d_0 = \dfrac{1}{h(z_0)} = \dfrac{1}{f'(z_0)}$. Hence for $z \in D \setminus \{z_0\}$,

$$
\frac{1}{f(z)} = \frac{1}{z - z_0} \cdot (d_0 + d_1(z - z_0) + \cdots) = \frac{d_0}{z - z_0} + d_1 + d_2(z - z_0) + \cdots
$$

and so $\operatorname{res}\left(\dfrac{1}{f}, z_0\right) = d_0 = \dfrac{1}{f'(z_0)}$.

Solution to Exercise 4.45

Let f be given by $f(z) = \sin z$. f has zeros of order 1 at $k\pi$, $k \in \mathbb{Z}$. So by the previous exercise,

$$\operatorname{res}\left(\frac{1}{\sin z}, k\pi\right) = \frac{1}{\sin' z|_{z=k\pi}} = \frac{1}{\cos(k\pi)} = \frac{1}{(-1)^k} = (-1)^k.$$

Solution to Exercise 4.46

(1) We have $f_0 = 1 \leq 2^0 = 1$, $f_1 = 1 \leq 2^1 = 2$, and if $f_m \leq 2^m$ for all $m \leq n$ (for some $n \geq 1$), then
$$f_{m+1} = f_m + f_{m-1} \leq 2^m + 2^{m-1} = 2^{m-1} \cdot 3 < 2^{m-1} \cdot 4 = 2^{m+1}.$$

(2) If $|z| < 1/2$, then $\sqrt[n]{|c_n z^n|} = \sqrt[n]{|c_n|} \cdot |z| \leq \sqrt[n]{2^n} \cdot |z| = 2|z| < 1$ for all $n \in \mathbb{N}$. So by the Root Test,
$$\sum_{n=0}^{\infty} |c_n z^n|$$
converges when $|z| < 1/2$. Hence the radius of convergence of F is $\geq 1/2$.

(3) We have for $|z| < 1/2$:
$$zF(z) = f_0 z + f_1 z^2 + f_2 z^3 + \cdots,$$
$$z^2 F(z) = \qquad f_0 z^2 + f_1 z^3 + \cdots.$$

Adding these,
$$zF(z) + z^2 F(z) = 1 \cdot z + (f_1 + f_0)z^2 + (f_2 + f_1)z^3 + \cdots$$
$$= f_1 z + f_2 z^2 + f_3 z^3 + \cdots$$
$$= (f_0 + f_1 z + f_2 z^2 + f_3 z^3 + \cdots) - f_0$$
$$= F(z) - 1.$$

Hence
$$1 = F(z) - zF(z) - z^2 F(z) = (1 - z - z^2)F(z).$$
So for $|z| < \dfrac{1}{2}$, we have $F(z) = \dfrac{1}{1 - z - z^2}$.

(4) We have
$$\frac{1}{z^{n+1}(1 - z - z^2)} = \frac{F(z)}{z^{n+1}}$$
$$= \frac{f_0 + \cdots + f_{n-1}z^{n-1} + f_n z^n + f_{n+1}z^{n+1} + \cdots}{z^{n+1}}$$
$$= \frac{f_0}{z^{n+1}} + \frac{f_1}{z^n} + \cdots + \frac{f_n}{z} + f_{n+1} + f_{n+2}z + \cdots,$$
$$(5.27)$$

and so

$$\text{res}\left(\frac{1}{z^{n+1}(1-z-z^2)}, 0\right) = \text{coefficient of } \frac{1}{z} \text{ in } (5.27) = f_n.$$

(5) We have for $|z| = R > 2$:

$$|1 - z - z^2| \geq |z^2 + z| - 1 = |z| \cdot |z + 1| - 1 = R \cdot |z + 1| - 1$$
$$\geq R \cdot (|z| - 1) - 1 = R \cdot (R - 1) - 1 = R^2 - R - 1$$
$$> 0 \quad (\text{since } R > 2).$$

Hence if $C_R : [0, 2\pi] \to \mathbb{C}$ is the circular path given by

$$C_R(t) = R\exp(it),$$

$t \in [0, 2\pi]$, then

$$\left|\int_{C_R} \frac{1}{z^{n+1}(1-z-z^2)}dz\right| \leq \frac{1}{R^{n+1}} \cdot \frac{1}{R^2 - R - 1} \cdot 2\pi R$$
$$= \frac{1}{R^n} \cdot \frac{1}{R^2 - R - 1} \xrightarrow{R \to \infty} 0.$$

Define G by $G(z) := \dfrac{1}{z^{n+1}(1-z-z^2)}$. Then G has

(a) a pole at 0 of order $n + 1$,

(b) a pole at $\dfrac{-1 + \sqrt{5}}{2}$ of order 1,

(c) a pole at $\dfrac{-1 - \sqrt{5}}{2}$ of order 1.

Thus

$$\text{res}(G, 0) + \text{res}\left(G, \frac{-1+\sqrt{5}}{2}\right) + \text{res}\left(G, \frac{-1-\sqrt{5}}{2}\right) = \frac{1}{2\pi i}\int_{C_R} G(z)dz$$

for all $R > 2$. Hence

$$\text{res}(G, 0) + \text{res}\left(G, \frac{-1+\sqrt{5}}{2}\right) + \text{res}\left(G, \frac{-1-\sqrt{5}}{2}\right)$$
$$= \lim_{R \to 0} \frac{1}{2\pi i}\int_{C_R} G(z)dz = 0,$$

that is, $f_n = -\text{res}\left(G, \dfrac{-1+\sqrt{5}}{2}\right) - \text{res}\left(G, \dfrac{-1-\sqrt{5}}{2}\right).$

We have

$$\mathrm{res}\left(G, \frac{-1+\sqrt{5}}{2}\right) = \lim_{z \to \frac{-1+\sqrt{5}}{2}} \left(z - \frac{-1+\sqrt{5}}{2}\right) \cdot \frac{1}{z^{n+1}(1 - z - z^2)}$$

$$= \frac{1}{\left(\dfrac{-1+\sqrt{5}}{2}\right)^{n+1} \cdot (-\sqrt{5})}$$

$$= \left(\frac{1+\sqrt{5}}{2}\right)^{n+1}\left(-\frac{1}{\sqrt{5}}\right).$$

Also,

$$\mathrm{res}\left(G, \frac{-1-\sqrt{5}}{2}\right) = \frac{1}{\left(\dfrac{-1-\sqrt{5}}{2}\right)^{n+1} \cdot \sqrt{5}} = \left(\frac{1-\sqrt{5}}{2}\right)^{n+1}\left(\frac{1}{\sqrt{5}}\right).$$

Hence

$$f_n = \frac{1}{\sqrt{5}} \cdot \left(\frac{1+\sqrt{5}}{2}\right)^{n+1} - \frac{1}{\sqrt{5}} \cdot \left(\frac{1-\sqrt{5}}{2}\right)^{n+1}$$

$$= \frac{1}{\sqrt{5}} \cdot \left(\left(\frac{1+\sqrt{5}}{2}\right)^{n+1} - \left(\frac{1-\sqrt{5}}{2}\right)^{n+1}\right).$$

Solutions to the exercises from Chapter 5

Solution to Exercise 5.1

(1) We have for $(x, y) \in \mathbb{R}^2 \setminus \{(0,0)\}$ that

$$\frac{\partial u}{\partial x} = \frac{1}{x^2 + y^2} \cdot 2x = \frac{2x}{x^2 + y^2},$$

$$\frac{\partial^2 u}{\partial x^2} = \frac{2}{x^2 + y^2} - \frac{2x}{(x^2 + y^2)^2} \cdot (2x) = \frac{2y^2 + 2x^2 - 4x^2}{(x^2 + y^2)^2} = \frac{2(y^2 - x^2)}{(x^2 + y^2)^2}.$$

Similarly noticing the symmetry in the roles of x and y, we have

$$\frac{\partial u}{\partial y} = \frac{2y}{x^2 + y^2}, \text{ and } \frac{\partial^2 u}{\partial y^2} = \frac{2(x^2 - y^2)}{(x^2 + y^2)^2}.$$

Consequently

$$\frac{\partial^2 u}{\partial x^2} + \frac{\partial^2 u}{\partial y^2} = \frac{2(y^2 - x^2)}{(x^2 + y^2)^2} + \frac{2(x^2 - y^2)}{(x^2 + y^2)^2} = 0.$$

Since u is C^2 and $\Delta u = 0$ in $\mathbb{R}^2 \setminus \{(0,0)\}$, u is harmonic there.

(2) We have for $(x, y) \in \mathbb{R}^2$ that

$$\frac{\partial u}{\partial x} = e^x \sin y, \quad \frac{\partial^2 u}{\partial x^2} = e^x \sin y, \text{ and,}$$

$$\frac{\partial u}{\partial y} = e^x \cos y, \quad \frac{\partial^2 u}{\partial y^2} = e^x(-\sin y).$$

So $\dfrac{\partial^2 u}{\partial x^2} + \dfrac{\partial^2 u}{\partial y^2} = e^x \sin y + e^x(-\sin y) = 0$. As u is C^2 and $\Delta u = 0$ in $\mathbb{R}^2$, u is harmonic in $\mathbb{R}^2$.

Solution to Exercise 5.2

Consider the vector space V of all real-valued functions defined on U, with pointwise operations. Then we know that V is a real vector space. We will show that $\mathrm{Har}(U)$ is a subspace of this vector space V, and hence a vector space with pointwise operations. We have

(S1) The constant function $\mathbf{0}$ assuming value 0 everywhere on U belongs to $\mathrm{Har}(U)$. Indeed

$$\frac{\partial^2 \mathbf{0}}{\partial x^2} + \frac{\partial^2 \mathbf{0}}{\partial y^2} = 0 + 0 = 0.$$

(S2) Let $u, v \in \mathrm{Har}(U)$. Then

$$\frac{\partial^2(u+v)}{\partial x^2} + \frac{\partial^2(u+v)}{\partial y^2} = \frac{\partial^2 u}{\partial x^2} + \frac{\partial^2 v}{\partial x^2} + \frac{\partial^2 u}{\partial y^2} + \frac{\partial^2 v}{\partial y^2}$$

$$= \left(\frac{\partial^2 u}{\partial x^2} + \frac{\partial^2 u}{\partial y^2}\right) + \left(\frac{\partial^2 v}{\partial x^2} + \frac{\partial^2 v}{\partial y^2}\right)$$

$$= 0 + 0 = 0.$$

(S3) Let $\alpha \in \mathbb{R}$ and $u \in \mathrm{Har}(U)$. Then

$$\frac{\partial^2(\alpha \cdot u)}{\partial x^2} + \frac{\partial^2(\alpha \cdot u)}{\partial y^2} = \alpha \cdot \frac{\partial^2 u}{\partial x^2} + \alpha \cdot \frac{\partial^2 u}{\partial y^2}$$

$$= \alpha \left(\frac{\partial^2 u}{\partial x^2} + \frac{\partial^2 u}{\partial y^2}\right) = \alpha \cdot 0 = 0.$$

Hence $\mathrm{Har}(U)$ is a real vector space with pointwise operations.

Solution to Exercise 5.3

$u := x = \mathrm{Re}(z)$ and $\tilde{u} := x + y = \mathrm{Re}(z - iz)$ are harmonic in $\mathbb{R}^2$, and their pointwise product is $u \cdot \tilde{u} = x \cdot (x + y) = x^2 + xy$. We have

$$\frac{\partial^2(u \cdot \tilde{u})}{\partial x^2} + \frac{\partial^2(u \cdot \tilde{u})}{\partial y^2} = \frac{\partial}{\partial x}(2x + y) + \frac{\partial}{\partial y}(x) = 2 + 0 = 2 \neq 0.$$

So the pointwise product of two harmonic functions need not be harmonic.

Solution to Exercise 5.4

(1) Let $u = e^x \sin y$. We seek a v such that $u + iv$ is holomorphic. So the Cauchy-Riemann equations must be satisfied. Hence

$$\frac{\partial v}{\partial x} = -\frac{\partial u}{\partial y} = -e^x \cos y.$$

So if we keep y fixed, we obtain by integrating that $v = -e^x \cos y + C(y)$, for a constant $C(y)$, which depends on y. Thus

$$\frac{\partial v}{\partial y} = e^x \sin y + C'(y) = \frac{\partial u}{\partial x} = e^x \sin y,$$

and so $C'(y) = 0$, giving $C(y) = K$. So we try $v := -e^x \cos y$. Then

$$u + iv = e^x \sin y + i(-e^x \cos y) = e^x(\sin y - i \cos y)$$

$$= -ie^x(\cos y + i \sin y) = -i \exp(x + iy) = -i \exp(z),$$

where $z = x + iy$. Hence $u + iv = -i \exp z$, which is indeed holomorphic. Hence $v = -e^x \cos y$ is a harmonic conjugate for $u := e^x \sin y$.

(2) Let $u = x^3 - 3xy^2 - 2y$. We seek a v such that $u + iv$ is holomorphic. So the Cauchy-Riemann equations must be satisfied. Hence

$$\frac{\partial v}{\partial x} = -\frac{\partial u}{\partial y} = 6xy + 2.$$

So if we keep y fixed, we obtain by integrating that

$$v = 6\frac{x^2}{2}y + 2x + C(y) = 3x^2y + 2x + C(y),$$

for a constant $C(y)$, which depends on y. Thus

$$\frac{\partial v}{\partial y} = 3x^2 + C'(y) = \frac{\partial u}{\partial x} = 3x^2 - 3y^2$$

and so $C'(y) = -3y^2$, which gives

$$C(y) = -3\frac{y^3}{3} + C = -y^3 + C,$$

and so we try $v = 3x^2y + 2x - y^3$. With this v, we have

$$u + iv = x^3 - 3xy^2 - 2y + i(3x^2y + 2x - y^3)$$
$$= x^3 + 3x(iy)^2 + 3x^2(iy) + (iy)^3 - 2y + i2x$$
$$= (x + iy)^3 + 2i(x + iy) = z^2 + 2iz$$

for $z = x + iy$. Thus $u + iv = z^2 + 2iz$, which is indeed holomorphic. So $v := 3x^2y + 2x - y^3$ is a harmonic conjugate of $u := x^3 - 3xy^2 - 2y$.

(3) Let $u := x(1 + 2y)$. We seek a v such that $u + iv$ is holomorphic. So the Cauchy-Riemann equations must be satisfied. Hence

$$\frac{\partial v}{\partial x} = -\frac{\partial u}{\partial y} = -2x.$$

So if we keep y fixed, we obtain by integrating that

$$v = -2\frac{x^2}{2} + C(y) = -x^2 + C(y),$$

for a constant $C(y)$, which depends on y. Thus

$$\frac{\partial v}{\partial y} = C'(y) = \frac{\partial u}{\partial x} = 1 + 2y.$$

Thus

$$C(y) = y + 2 \cdot \frac{y^2}{2} + C = y + y^2 + C.$$

So we try $v := -x^2 + y + y^2$. With this v, we have

$$u + iv = x(1 + 2y) + i(-x^2 + y + y^2) = x + iy + 2xy + i(y^2 - x^2)$$
$$= x + iy - i((x^2 - y^2) + i2xy) = x + iy - i(x + iy)^2$$
$$= z - iz^2$$

for $z = x + iy$. Hence $u + iv = z - iz^2$ is indeed holomorphic, and $v := -x^2 + y + y^2$ is a harmonic conjugate of $u := x(1 + 2y)$.

Solution to Exercise 5.5

Let v be a harmonic conjugate of u. Then $f := u + iv$ is holomorphic in $\mathbb{C} \setminus \{0\}$. Hence $h := z^2 \exp(-f(z))$ is holomorphic in $\mathbb{C} \setminus \{0\}$. We have

$$|h| = |z|^2 |\exp(-f(z))| = |z|^2 e^{-\mathrm{Re}(f(z))} = |z|^2 e^{-u} = |z|^2 e^{-\log|z|^2}$$

$$= |z|^2 \cdot \frac{1}{|z|^2} = 1.$$

But then h must be constant in each disc contained in $\mathbb{C} \setminus \{0\}$. Consequently h must be constant in $\mathbb{C} \setminus \{0\}$. Thus $h' = 0$. But

$$h' = 2z \exp(-f(z)) + z^2 \exp(-f(z)) \cdot (-f'(z)),$$

and so

$$f'(z) = \frac{2}{z}.$$

So $1/z$ would have a primitive in $\mathbb{C} \setminus \{0\}$. Now if γ is the path $\gamma(t) = \exp(it)$, $0 \le t \le 2\pi$, we have the contradiction that

$$2 \cdot 2\pi i = \int_\gamma \frac{2}{z} dz = \int_\gamma f'(z) dz = 0.$$

Hence u has no harmonic conjugate in $\mathbb{C} \setminus \{0\}$.

Solution to Exercise 5.6

Set $u := x^3 + y^3$. If f were holomorphic, then u would be harmonic. But

$$\frac{\partial^2 u}{\partial x^2} + \frac{\partial^2 u}{\partial y^2} = \frac{\partial}{\partial x}(3x^2) + \frac{\partial}{\partial y}(3y^2) = 6x + 6y = 6(x+y) \ne 0$$

for $x \ne -y$. Hence the answer is "no".

Solution to Exercise 5.7

It suffices to show that if u is harmonic, then so are $\dfrac{\partial u}{\partial x}$ and $\dfrac{\partial u}{\partial y}$.

We know that u is infinitely many times differentiable. We have

$$\frac{\partial^2}{\partial x^2}\left(\frac{\partial u}{\partial x}\right) + \frac{\partial^2}{\partial y^2}\left(\frac{\partial u}{\partial x}\right) = \frac{\partial}{\partial x}\left(\frac{\partial^2 u}{\partial x^2}\right) + \frac{\partial}{\partial y}\left(\frac{\partial^2 u}{\partial y \partial x}\right)$$

$$= \frac{\partial}{\partial x}\left(-\frac{\partial^2 u}{\partial y^2}\right) + \frac{\partial}{\partial y}\left(\frac{\partial^2 u}{\partial x \partial y}\right)$$

$$= \frac{\partial}{\partial x}\left(-\frac{\partial^2 u}{\partial y^2}\right) + \frac{\partial}{\partial x}\left(\frac{\partial}{\partial y}\left(\frac{\partial}{\partial y} u\right)\right)$$

$$= \frac{\partial}{\partial x}\left(-\frac{\partial^2 u}{\partial y^2} + \frac{\partial^2 u}{\partial y^2}\right) = \frac{\partial}{\partial x}(0) = 0.$$

Similarly

$$\frac{\partial^2}{\partial x^2}\left(\frac{\partial u}{\partial y}\right) + \frac{\partial^2}{\partial y^2}\left(\frac{\partial u}{\partial y}\right) = \frac{\partial}{\partial y}\left(\frac{\partial^2 u}{\partial x^2}\right) + \frac{\partial}{\partial y}\left(\frac{\partial^2 u}{\partial y^2}\right)$$

$$= \frac{\partial}{\partial y}\left(\frac{\partial^2 u}{\partial x^2} + \frac{\partial^2 u}{\partial y^2}\right) = \frac{\partial}{\partial y}(0) = 0.$$

Solution to Exercise 5.8

(1) Let $b(x) = p(x) = c_0 + c_1 x + \cdots + c_d x^d$. Then

$$p(z) := p(x + iy) = c_0 + c_1 z + \cdots + c_d z^d$$

is entire, and so $h := \mathrm{Re}(p(x+iy))$ is harmonic. Moreover for all $x \in \mathbb{R}$,

$$h(x, 0) = \mathrm{Re}(p(x + i0)) = \mathrm{Re}(p(x)) = \mathrm{Re}(b(x)) = b(x).$$

(2) We have $b(z) := b(x + iy) := \dfrac{1}{1 + z^2}$ is not defined at $z = i$. But

$$\frac{i}{z + i}$$

is holomorphic in the upper half-plane, and so its real part is harmonic there. Moreover,

$$h(x, 0) = \frac{0 + 1}{x^2 + (0 + 1)^2} = \frac{1}{x^2 + 1} = b(x)$$

for all $x \in \mathbb{R}$.

Solution to Exercise 5.9

Let f be an entire function whose real part is u. (Note that $\mathbb{C}$ is simply connected.) Then $\exp(-f)$ is entire too. We have

$$|\exp(-f)| = e^{-\mathrm{Re}(f)} = e^{-u} \leq 1,$$

since $u(x, y) > 0$ for all $x, y \in \mathbb{R}$. By Liouville's Theorem, $\exp(-f)$ is a constant. Hence $|\exp(-f)|$ is constant too, that is, e^{-u} is constant. Consequently the real logarithm $\log(e^{-u}) = -u$ is constant, and so u is constant as well.

Solution to Exercise 5.10

(1) For $z = r\exp(i\theta)$, we have
$$\exp\left(-\frac{1}{z^4}\right) = \exp\left(-\frac{1}{r^4}\exp(-i4\theta)\right).$$
So if we take $r = 1/n$ and $4\theta = -\pi$, that is, with
$$z_n := \frac{1}{n}\exp\left(-i\frac{\pi}{4}\right) =: x_n + iy_n,$$
we have $u(x_n, y_n) = \exp(-n^4\exp(i\pi)) = \exp(-n^4\cdot(-1)) = e^{n^4}$. So we have that $(x_n, y_n) \to (0, 0)$, but it is not the case that $u(x_n, y_n) \to 0$, showing that u is not continuous at $(0, 0)$.

(2) We have
$$u(x, 0) = \exp\left(-\frac{1}{(x + 0i)^4}\right) = e^{-1/x^4}, \text{ and}$$
$$u(0, y) = \exp\left(-\frac{1}{(0 + yi)^4}\right) = \exp\left(-\frac{1}{i^4y^4}\right) = \exp\left(-\frac{1}{1\cdot y^4}\right) = e^{-1/y^4}.$$

(3) We have
$$\frac{\partial u}{\partial x}(0, 0) = \lim_{x\to 0}\frac{u(x, 0) - u(0, 0)}{x - 0} = \lim_{x\to 0}\frac{e^{-1/x^4} - 0}{x} = \lim_{x\to 0}\frac{e^{-1/x^4}}{x} = 0.$$
(The last equality follows from the fact that
$$e^{1/x^4} = 1 + \frac{1}{x^4} + \frac{1}{2!}\left(\frac{1}{x^4}\right)^2 + \cdots > \frac{1}{x^4}$$
and so $0 \leq \left|\dfrac{e^{-1/x^4}}{x}\right| \leq |x|^3$.) Similarly, $\dfrac{\partial u}{\partial y}(0, 0) = 0$. Thus
$$\frac{\partial^2 u}{\partial x^2}(0, 0) = \lim_{x\to 0}\frac{\frac{\partial u}{\partial x}(x, 0) - \frac{\partial u}{\partial x}(0, 0)}{x - 0} = \lim_{x\to 0}\frac{\frac{d}{dx}e^{-1/x^4} - 0}{x}$$
$$= \lim_{x\to 0}\frac{e^{-1/x^4}\cdot\frac{-4}{x^5}}{x} = \lim_{x\to 0}\frac{-4e^{-1/x^4}}{x^6} = 0.$$
(As $e^{1/x^4} = 1 + \dfrac{1}{x^4} + \dfrac{1}{2!}\left(\dfrac{1}{x^4}\right)^2 + \cdots > \dfrac{1}{2x^8}$, $0 \leq \left|\dfrac{e^{-1/x^4}}{x^6}\right| \leq 2|x|^2$.)
Similarly,
$$\frac{\partial^2 u}{\partial y^2}(0, 0) = 0.$$
Hence $\dfrac{\partial^2 u}{\partial x^2}(0, 0) + \dfrac{\partial^2 u}{\partial y^2}(0, 0) = 0 + 0 = 0.$

Solution to Exercise 5.11

(1) Let $z_0 \in D_1$. Then $\varphi(z_0) \in D_2$. Let Δ be a disc with center $\varphi(z_0)$ and radius $\epsilon > 0$ small enough so that $\Delta \subset D_2$ and $\varphi^{-1}(\Delta) \subset D_1$. Since Δ is simply connected, there is a holomorphic function G defined in Δ such that $g = \operatorname{Re}(G)$ in Δ. The composition of the holomorphic maps $\varphi|_{\varphi^{-1}(\Delta)} : \varphi^{-1}(\Delta) \to \Delta$ and $G : \Delta \to \mathbb{C}$ is holomorphic, and so $\operatorname{Re}(G \circ \varphi|_{\varphi^{-1}(\Delta)})$ is harmonic in $\varphi^{-1}(\Delta)$. But for $z \in \varphi^{-1}(\Delta)$, $\varphi|_{\varphi^{-1}(\Delta)}(z) = \varphi(z) \in \Delta$, and so

$$(G \circ \varphi|_{\varphi^{-1}(\Delta)})(z) = G(\varphi(z)) = g(\varphi(z)) = (g \circ \varphi)(z).$$

So $(g \circ \varphi)|_{\varphi^{-1}(\Delta)}$ is harmonic in $\varphi^{-1}(\Delta)$. As $z_0 \in \varphi^{-1}(\Delta)$, in particular, we have $\Delta(g \circ \varphi)(z_0) = 0$. Since the choice of $z_0 \in D_1$ was arbitrary, $g \circ \varphi$ is harmonic in D_1.

(2) If $h : D_2 \to \mathbb{R}$ is harmonic, then by the first part, it follows that $h \circ \varphi : D_1 \to \mathbb{R}$ is harmonic.

Now suppose $h \circ \varphi : D_1 \to \mathbb{R}$ is harmonic. Then since $\varphi^{-1} : D_2 \to D_1$ is holomorphic, by the first part, $(h \circ \varphi) \circ \varphi^{-1} : D_2 \to \mathbb{R}$ is harmonic. But $(h \circ \varphi) \circ \varphi^{-1} = h \circ (\varphi \circ \varphi^{-1}) = h \circ (\mathrm{id}_{D_2}) = h$, where $\mathrm{id}_{D_2} : D_2 \to D_2$ is the identity map $z \mapsto z$ ($z \in D_2$). So $h : D_2 \to \mathbb{R}$ is harmonic.

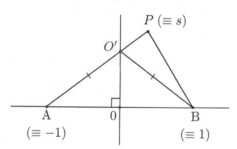

Fig. 5.24 Triangle inequality in $\triangle PO'B$

(3) By the triangle inequality in $\triangle PO'B$ shown in Figure 5.24, for $s \; (\equiv P)$ in $\mathbb{H}$, we have

$$|s + 1| = \ell(PA) = \ell(PO') + \ell(O'A) = \ell(PO') + \ell(O'B)$$
$$> \ell(PB) = |s - 1|,$$

where we have used the fact that O' is on the perpendicular bisector of AB to get the third equality. So $\varphi(s) \in \mathbb{D}$ for all $s \in \mathbb{H}$. The function φ is clearly holomorphic: for $s \in \mathbb{H}$,

$$\varphi'(s) = 1 \cdot \frac{1}{s+1} + (s - 1) \cdot \left(-\frac{1}{(s+1)^2} \right) = \frac{s + 1 - s + 1}{(s+1)^2} = \frac{2}{(s+1)^2}.$$

Now consider $\psi : \mathbb{D} \to \mathbb{H}$ given by

$$\psi(s) = \frac{1+z}{1-z}, \quad z \in \mathbb{D}.$$

(This expression for ψ, which is a candidate for φ^{-1}, is obtained by solving for s in the equation $z = \varphi(s) = \dfrac{s-1}{s+1}$.)

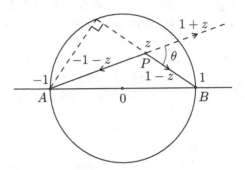

Fig. 5.25   Triangle inequality in $\triangle PO'B$

In the Figure 5.25, we know that the angle subtended by the diameter AB at any point of the circle is $90°$, and so for any point $P \ (\equiv z)$ in $\mathbb{D}$, $\angle APB > 90°$. So

$$\mathrm{Re}(\psi(z)) = \mathrm{Re}\left(\frac{1+z}{1-z}\right) = |\psi(z)| \cos\theta = |\psi(z)| \cos(\pi - \angle APB) > 0.$$

Thus $\psi(z) \in \mathbb{H}$ for all $z \in \mathbb{D}$. The map ψ is holomorphic in $\mathbb{D}$: for $z \in \mathbb{D}$,

$$\psi'(z) = 1 \cdot \frac{1}{1-z} + (1+z) \cdot \left(\frac{1}{(1-z)^2}\right) = \frac{1-z+1+z}{(1-z)^2} = \frac{2}{(1-z)^2}.$$

Finally, for $s \in \mathbb{H}$, we have

$$(\psi \circ \varphi)(s) = \frac{1 + \dfrac{s-1}{s+1}}{1 - \dfrac{s-1}{s+1}} = \frac{s+1+s-1}{s+1-s+1} = \frac{2s}{2} = s,$$

and for $z \in \mathbb{D}$, we have

$$(\varphi \circ \psi)(z) = \frac{\dfrac{1+z}{1-z} - 1}{\dfrac{1+z}{1-z} + 1} = \frac{1+z-1+z}{1+z+1-z} = \frac{2z}{2} = z.$$

So φ is a bijection and $\varphi^{-1} = \psi$.

Bibliography

Beck, M., Marchesi, G., Pixton, D., and Sabalka, L. (2008). *A First Course in Complex Analysis*, http://math.sfsu.edu/beck/papers/complex.pdf .

Conway, J. (1978). *Functions of One Complex Variable I*, 2nd Edition. (Springer).

Fisher, S. (1999). *Complex Variables*, 2nd Edition. (Dover).

Flanigan, F. (1972) *Complex Variables* (Dover).

Flanigan, F. (1973). Classroom Notes: Some Half-Plane Dirichlet Problems: A Bare Hands Approach. *American Mathematical Monthly* **80**, 1, pp. 59-61.

Gelbaum, B. and Olmsted, J. (1964). *Counterexamples in Analysis* (Dover).

Gilman, J., Kra, I. and Rodriguez, R. (2007). *Complex Analysis. In the Spirit of Lipman Bers* (Springer).

Howie, J. (2003). *Complex Analysis* (Springer).

Needham, T. (1997). *Visual Complex Analysis* (Oxford University Press).

Ash, R. and Novinger, W. (2007). *Complex Analysis*, 2nd Edition. (Dover).

Remmert, R. (1991) *Theory of Complex Functions* (Springer).

Rudin, W. (1987). *Real and Complex Analysis*, 3rd edition. (McGraw-Hill).

Shastri, A. (2000). *An Introduction to Complex Analysis* (Macmillan Publishers India).

Shaw, W. (2006). *Complex Analysis with MATHEMATICA* (Cambridge University Press).

Shurman, J. (2012). *Course Materials for Mathematics 311: Complex Analysis*, http://people.reed.edu/~jerry/311/mats.html .

Tall, D. (1970) *Functions of a Complex Variable* (Dover).

Volkovyskiĭ, L., Lunts, G. and Aramanovich, I. (1991). *A Collection of Problems on Complex Analysis* (Dover).

Index